Sensuous Seas

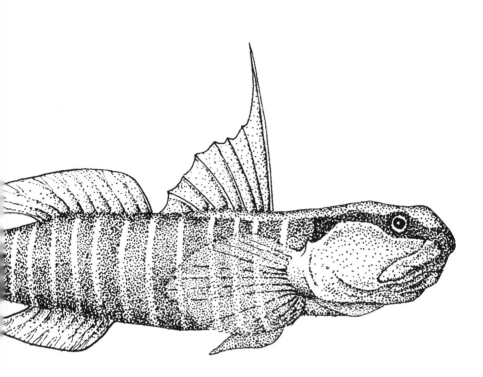

EUGENE H. KAPLAN

*Drawings by Sandy Chichester Rivkin
and Susan L. Kaplan*

Sensuous Seas

Tales of a Marine Biologist ❧

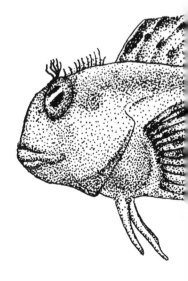

Princeton University Press
Princeton and Oxford

Published by Princeton University Press, 41 William Street, Princeton, New Jersey 08540

In the United Kingdom: Princeton University Press, 3 Market Place, Woodstock, Oxfordshire OX20 1SY

Library of Congress Cataloging-in-Publication Data

Kaplan, Eugene H. (Eugene Herbert), 1932–
Sensuous seas : tales of a marine biologist / Eugene H. Kaplan.
p.   cm.
Includes bibliographical references (p.).
ISBN-13: 978-0-691-12560-2 (cloth : alk. paper)
ISBN-10: 0-691-12560-0 (cloth : alk. paper)
1. Marine biology.   2. Sexual behavior in animals.   3. Marine biology—Anecdotes.
4. Sexual behavior in animals—Anecdotes.   5. Kaplan, Eugene H. (Eugene Herbert), 1932–
I. Title.
QH91.1.K37 2006
578.77—dc22   2005054517

British Library Cataloging-in-Publication Data is available

This book has been composed in Fournier with Caslon Open Face display

Printed on acid-free paper.

pup.princeton.edu

Printed in the United States of America

10 9 8 7 6 5 4 3 2 1

❧❧❧ CONTENTS

What Is a Marine Biologist?

WHO IS THE MOST FAMOUS George of all time? Not George Washington, he of the wooden dentures; not George Steinbrenner, of flinty disposition. No, not either of these esteemed gentlemen. The most famous is the unfortunate George of Seinfeld fame. George's glory was that he was the greatest loser of all time. Millions sat glued to their TVs, anxious to help George out of his latest disaster.

One episode begins with Kramer hitting golf balls into the sea. Later George is seen standing on the beach. "What will be the denouement of this oddly innocuous juxtaposition of activities?" we asked ourselves. There, on the horizon, appears the inevitable, inaccessible beautiful lady. She approaches, they talk. Groping for the most heroic identity he can come up with, George tells her, "I am a marine biologist." As he says this, a crowd gathers—there is a huge whale stranded just offshore!

The young lady says to George, "You are a marine biologist—do something." George's bluff is called. He looks stricken, then determined. He rolls up his pants and wades out into the water. The damsel looks on with fear in her eyes—then joy. George returns from the sea, holding his hand up in triumph. He has saved the whale. In his hand is Kramer's golf ball. It had been lodged in the blowhole of the whale, suffocating it.*

George's adventure notwithstanding, there is no such thing as a "marine biologist." One can be a marine invertebrate zoologist (worms, clams, and crabs), a marine ichthyologist (fishes), or even a marine phycologist (algae, seaweed). Within these specialties one can study the ecology, behavior, physiology, taxonomy, and so on of one's chosen group of organisms. The

---

* Many reviewers have corrected my distorted version of this story. It seems to me that everyone has memorized this episode. Because, in this rare instance, I am permitted poetic license, there will be no attempt to reach perfection. I prefer my dimly remembered synopsis. It makes more sense to me.

marine scientist usually dedicates his life to the study of one kind of organism, like sharks, and one kind of intellectual quest, like tracing the migration patterns of different sharks.

Each year I am deluged with phone calls from young people who want to become "marine biologists," mostly marine mammal trainers. This is television's influence. Animal training has become synonymous with marine biology.

To my amazement, I met a young woman who had quit this much-desired position at our local public aquarium. Incredulous, I asked her why she gave up this plum of a job. She answered, "The walrus was always snotting on me."

THIS BOOK is dedicated to the memory of Robert W. Johnson, PhD, Distinguished Professor of Conservation and Ecology at Hofstra University, friend, comrade, advisor, and fellow fisherman. He was the only individual I know who lived up to the oxymoron "compassionate conservative."

My inspiration for this book came from the literally thousands of students whose somnolent eyes challenged me to stretch to the utmost to convey my fascination with the underwater world. Many participated in my field courses in Jamaica and the British Virgin Islands, slogging through swamps and snorkeling over coral reefs. Outstanding among these students are professors Paul Billeter and John Morrissey—friends, colleagues, and, I hope, disciples. They have fun teaching watery subjects, and I see my happiest years reflected in their classes.

My deepest appreciation goes to my long-suffering assistant, Debra Bradley.

Never ending support, stability, and patience were provided by my wife, Breena and daughters, Julie and Sue. Their love made this book possible. In addition to her familial contribution, daughter Susan Kaplan provided drawings for this book and for the two Peterson Field Guides.

Since its inception twenty-odd years ago, the Hofstra University Marine Laboratory in St. Anns Bay, Jamaica, has been a focus of my teaching. It was here that I was able to expose students to the fantastic underwater worlds of the coral reef and tropical shores. Thousands of students (and not a few senior citizens/elderhostelers) have been overwhelmed and flabbergasted by their learning experience at HUML.

It has been my privilege to work with the staff of HUML, all Jamaicans: Edgar Ross, Sandi Ross, Sandy Walters, Nolbert Campbell, Clive Richards, Augustus Keize, "Sharty," and the other workers.

My daughter Julie Kaplan administered HUML and kept it going, taking care of the infinite minutiae with competence and good humor.

Each year HUML was placed in the care of two scientists who supervised the educational/scientific programs. These resident directors lectured and led field trips where they dove, snorkeled, marched through swamps, clambered over sharp rocks, picked up vicious crabs, were bitten by fish, mosquitoes, and other animals, bandaged wounds, carried drunk students to their beds, and did all the myriad tasks necessary to run a field station. They were truly dedicated and an unusually competent bunch. Several have created clones of HUML; all the others have gone on to creative jobs—a testament to HUML's influence on their lives. My sincere gratitude goes to: Amy Edwards, John and Alice Morrissey, Drew and Dodie Ferrier, Tom and Leann Byrnes, Steve Vee, Bill Allison, Bill and Chrissy Janssen, Bruce and Brenda Adkinson, Mel and Gail Zimmerman, Wendy Lee and Ray Barneveld, Ken and Maureen Mattes, Andy and Robin Bruckner, Dave and Kelly Hutchinson, Ryan Cilsick and Michelle Anderson, Sarah Brondson and Magnus Johnson, Rick and Donna Nemeth, Deborah Gochfeld and Dwayne Minton, Silvia Macia and Mike Robinson, and Trish Hankenson and Mike Lewis.

Sandy Chichester Rivkin stepped in at a moment of crisis and created masterful drawings. Wonderful people at Princeton University Press shepherded this book through the challenges of production. I am indebted to Ellen Foos, Dimitri Karetnikov, Lyman Lyons, and Maria Lindenfeldar. Robert Kirk saw merit in this odd book when it was just an embryo. All have gone above and beyond the call of duty.

I very much appreciate the efforts of colleagues who read the whole manuscript in its formative stages for scientific validity. Professors Leland Pollock, Paul Billeter, and Joe Britton provided more than knowledge, voluntarily extending their mandate to providing much needed wise counsel.

All mistakes in this book are my sole responsibility.

ANTHROPOMORPHISM AND TELEOLOGY, two of the mad horsemen of the biological apocalypse, run rampant through the pages of this book. Anthropomorphism is to give human characteristics to animals or objects. ("The Little Engine That Could" is an example of starting children off on the wrong logical foot). For example, abundant references have been made to the passions of fiddler crabs and sea horses and blushing maiden octopods. The patently prurient nature of the stories herein seemed to require committing this sin to make them more interesting.

Teleology is confusing cause with effect. Any implication that an organism has evolved to acquire a biological niche (the giraffe evolved a long neck so that it could eat leaves off the tops of trees) is an example of poetic license. I have railed against these facets of illogical thinking in the field of biology in another book, and I apologize for resorting to them in the name of storytelling.

Sensuous Seas

The Perils of Teaching

I AM STANDING at the front of the classroom on the first day of class, wondering what inspirational message to deliver to create the proper learning environment. The class is quiet—few of the students are acquainted with one another as yet. The door opens and in walks Miss Nubile, the fabric of her white blouse straining in beautiful curves like the sails of a ship running before the wind. She is armed to the teeth, like that most fearsome of old time sailing battleships, the Portuguese man o' war. Her cannons are her luscious curves and there is an aura about her that makes the eye of each male slide toward her, and the mouth of each female in the class turn down. The stage is set for a showdown.

How am I to turn the thoughts of these hormone-laden young men and women to the subject matter at hand? What can I say that will interest them, I wonder, as I drone on interminably about clams and worms? Then I remember the advice my learned friend and distinguished professor of ecology whispered in the dark inner sanctum of the local bar, the secret of good biology teaching: *if you can't beat them, join them*. Infuse into each lecture a generous helping of sex, so that seething hormone-infused thoughts will be directed toward what I am saying, not Miss Nubile.

This book is a compendium of stories I used to "stimulate" students, stories of mating behavior and sexual habits of animals that live in the ocean. But I will not be compulsive about sex in the sea. I will throw in the bizarre, like the candiru, the only parasitic fish that attacks humans. It enters the penis, swims up the urethra, lodges in the urinary bladder, and chews its way through the walls of the bladder, engorging itself with blood until the host dies a horrible death. Visualize the student, paying rapt attention as I go on about the mundane clam, not wanting to miss the moment of digression— "*When will he get to the good stuff?*"

Let's get something straight: I did not deliberately place myself into dangerous underwater situations to enhance my teaching. It is inevitable

1

that near-drowning situations arise during forty years of field investigations and teaching. I did not, like my dedicated friend, deliberately walk into a cove containing schools of man-eating sharks to prove the validity of a hypothesis (see chapter 30). Self-sacrifice is beyond my motivational incentives. But scary stories of my near-drowning certainly enhance and enliven my lectures.

The analogy likening Miss Nubile to a Portuguese man o' war, used to introduce a subsequent subject, is an example of each chapter's structure: a dramatic motivational story followed by biological content enlivened by personal experiences.

1

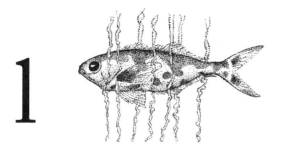

Deadly Darts

Brainless, boneless, bloodless . . . blobs . . .
successfully oceangoing for 650 million years.
—LILY WHITEMAN*

IRIDESCENT PURPLISH BALLOONS skittered across the sea in a fresh breeze,
destined to wash up in windrows on a sandy beach like the remnants of
a child's birthday party. A little boy wandered along. Curious, he bent over
to pick up a stranded "balloon." The "string" touched his leg. An excruciating pain emanated from the point of contact. The child staggered back and
fell among the balloons. He writhed in agony, the stings causing a spiderweb
of red welts like whiplashes on his skin. A stranger happened along and carried the now semiconscious child to the nearest first aid station, where his
body was slathered with meat tenderizer. The protein-destroying enzyme in
the tenderizer destroyed the toxin. The child survived after hospitalization.

Shaken after the horrific scene played out in front of me, I walked over
to the shore. Among the footprints of the saved and the savior were a few
of the balloons. Recognizing them for what they were, I looked around for

---

* Lily Whiteman, "The Blobs of Summer," *On Earth* (National Resources Defense
Council), Summer, 2002.

something to put one in. I needed a photo of the creature and recalled the dictum, "the best way to take an underwater photo is not to take it underwater." Back at my motel room was a water-filled lunch box-sized aquarium. A camera with macro lens stood poised on a tripod, pointed at a potential aquatic subject.

The beach was pristine—no washed-up plastic cups, no Coke bottles. How could I carry the specimen back to the "photography studio" I had set up in my room? Then I thought of the dive mask on my forehead. I scooped up a little water in the mask, slid it under a balloon, rushed back to the room, and plopped the specimen into the tiny tank. Its tentacles moved up and down. The balloon even writhed around, so that a sequence of photos would prove that this purple sphere was capable of movement.

About two hours later, after the photo session and lunch, I returned to the shore for more photo ops. I put on my mask. Suddenly my face was on fire! The pain was so intense that I gasped and ripped off the mask. My eyes were swollen nearly shut as I rushed back to the room. We had no meat tenderizer, but my wife applied a topical anesthetic.

After half an hour the haze of pain lifted and I was able to think. "What happened?" I asked myself. I realized that there were some tentacles left in the mask, and despite drying for hours in the sun, they still retained their toxicity.

❧❧❧❧

The innocent-looking balloons were in reality biological bladders filled with air secreted by dangling colonies of tiny elongate animals. The villain of the piece is the Portuguese man o' war, *Physalia physalia*, among the most fearsome of jellyfishes. Few biologists know the origin of the name "Portuguese man o' war." It was derived from a powerful four-masted battleship, bristling with at least thirty-eight cannons and characterized by two large, voluptuous lateen sails and a substantial stern (like Miss Nubile). This warship made it possible for the Portuguese to dominate the seas during the sixteenth century.

The ship's biological namesake is also formidable. This bizarre jellyfish is an animal of such simplicity that its functions are divided among three body types called polyps: one for defense, one for feeding, and one for reproduction. These quarter-inch-long, interconnected, semi-independent polyps dangle from the bottom of the balloon and combine their functions for the greater good. In other words, the phylum has not yet evolved a body that can perform all of the life functions. Like the Borg, each function is performed by a specific part of the collective body that is connected with the others. Food is eaten by a gastrozoid, reproduction is performed by a gonozoid, and protection is provided by a dactylozoid. In the case of the Portuguese man o' war, the defensive dactylozoids evolved to become the aggressive members of the triumvirate. They extend filamentous fishing tentacles twenty feet behind the floating colony, ensnaring passing fishes and zooplankton in an almost invisible web of toxic threads.

How has this phylum, the Cnidaria* so primitive as to lack organs and virtually just a jelly-filled sack, existed for 650 million years? No brain, no blood, no heart, no anus. Yet the phylum has survived fundamentally unchanged over the millenia, so it must have something going for it. That something is a poison arrow, the nematocyst. Each tentacle is covered with thousands of cells that are capable of discharging poisonous nematocysts in an explosion of toxicity. So tiny are these ancient weapons that in their coiled state they are scarcely larger than the nucleus of the cell. In typically huge numbers, the microscopic darts are capable of introducing considerable amounts of toxin into the superficial layer of the victim's skin. The toxin must be very powerful indeed if the small amount that penetrates the epidermis can cause humans to experience such intense pain and small fish such instantaneous paralysis. Visualize the hairs on your arm as poisonous weapons and you will have an idea of what an aquatic organism faces when it rubs against a tentacle.

The basic cnidarian life cycle consists of two independent reproductive forms, one sexual (the medusa) and the other asexual (the polyp). In one large group, including corals and sea anemones, the polyp incorporates the

---

* The phylum is called Cnidaria (kneè-daria) after the Greek word *cnidus*, meaning thread. The thread referred to is the tiny, painful, hair-like spine of a plant called the nettle, *Urtica*. To sit on a nettle plant is to be stung with tiny, needle-like "threads."

PLATE I

A. THE PORTUGUESE MAN O' WAR, *Physalia physalia*, is a 12-inch purple translucent bladder filled with nitrogen-rich air secreted by polyps dangling below. The colony has many 20-foot fishing tentacles. Some of these have retracted, pulling a paralyzed fish toward the rest of the colony suspended from the bladder, where hundreds of feeding polyps will digest the fish.

B. THE MAN O' WAR FISH, *Nomeus gronovii*, flourishes among the tentacles although vulnerable to their fatal sting. The fish maneuvers among the toxic tentacles and darts out to capture its planktonic food.

C. THE PORTUGUESE MAN O' WAR SHIP, with up to seventy-two cannons, was instrumental in maintaining the Portuguese navy's dominance of the seas in the fifteenth and sixteenth centuries. It evolved from a merchant ship, the caravel (exemplified by Columbus's *Nina*), into the galleon depicted. Note the two voluptuous triangular lateen sails near the stern.

D. TYPES OF TENTACLES. Those on right are male and female reproductive polyps, gonozoids. They "ripen" at different times, preventing self-fertilization. The central polyp, the dactylozoid, is a coiled, retractable fishing tentacle armed with fierce nematocysts in batteries. Three feeding polyps, gastrozoids, are to the left between two dactylozoids. When the dactylozoids pull the prey close to the colony, gastrozoids will extend and secrete enzymes to digest the prey.

E. A COPEPOD paralyzed by venomous nematocysts from a tentacle (those with bulbs torn from the battery). Sticky, whip-like nematocysts stay attached to the tentacle to hold the prey until the coiled tentacle retracts and carries it to the colony. Nematocysts are in spherical batteries in this species. In other species they are distributed like hairs on your arm. Each nematocyst bursts from a single cell.

PLATE I

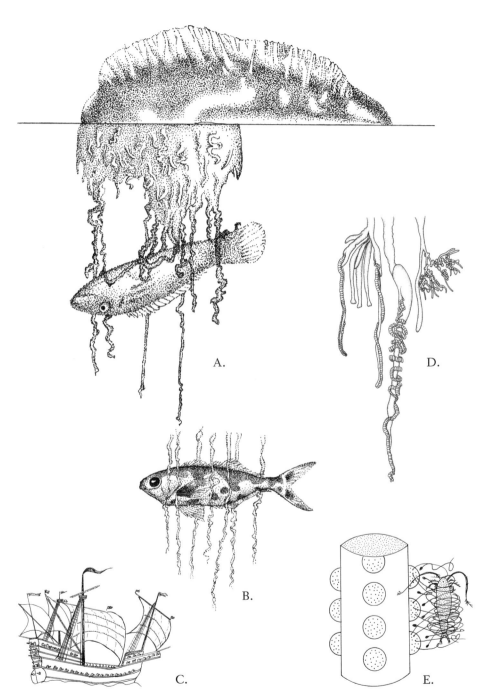

A.

D.

B.

C.

E.

sexual phase and there is no medusa. When a medusa is present, this sexual floating stage is popularly known as the jellyfish. Scientists coined the term "medusa" because it reminded them of the snake-headed mythical monster who turned men into stone when they looked at her. The poisonous, snake-like tentacles of the medusa literally turn a small fish into stone—total paralysis, so that the death shudder is suppressed. A medusa produces either eggs or sperm and casts them into the sea. They fuse to become the asexual polyp. Many polyps clone to form fuzz-like colonies attached to hard objects on the bottom. These colonies then bud off juvenile jellyfish.

The Portuguese man o' war differs from the typical cnidarian, being neither medusa nor polyp. It is a floating colony of polyps suspended from a bladder of its own making. The downward-pointing polyps, en masse, manufacture the purple balloon, injecting special nitrogen-rich air into it. Although the balloon can contort itself, swimming is impossible and the colony goes where it is blown.

The illustration depicts a Portuguese man o' war sailing majestically along, wafted by the wind, a beautiful purple air-filled sphere trailing its fierce weapons behind—twenty-foot-long, nematocyst-laden, transparent, string-like fishing tentacles. It has captured a small fish. But the fish might escape, tearing off thread-like tentacles with one convulsive movement, partly disarming the colony. To prevent this, the Portuguese man o' war must paralyze the fish instantaneously. After the prey is captured, the tentacle shortens, carrying its paralyzed victim to the tiny gaping mouths of the feeding polyps.

Nature, ever experimental, has come up with a surprisingly benevolent aspect to the fierce Portuguese man o' war. It provides a haven for the man o' war fish *Nomeus*, which finds protection among the malevolent nematocyst-bearing tentacles. What physiological mechanism has the fish evolved to foster this intimate relationship? Apparently none, for if the jellyfish is removed from the water and its resident fish falls on the tentacles, the fish is immediately paralyzed. But its survival depends on its ability to swim among the tentacles. Only one explanation is possible. *Nomeus* must have evolved an exquisitely sensitive sensory mechanism that allows it to live in a virtual web of danger and avoid getting stung.

2

The Great Jade Green Octopus Hunt

RAGGEDLY, IN A CONFUSED CADENCE, the words wafted across the velvet black night, "Oh say can you see by the dawn's early light?" Knee deep in the Caribbean Sea, surrounded by spiny sea urchins, a class of frightened bathing suit-clad American students was singing the stirring words of the Star Spangled Banner. In the distance a storm was brewing. The roar of thunder, punctuated by flashes of lightning, uncannily evoked the moment of creation of our national anthem in the "rockets' red glare."

We were standing in the warm sea about a quarter mile off the Jamaican shore. Dense darkness rendered the shoreline of palm trees and white beaches invisible. The light emanating from the students' flashlights created greenish transparent halos that punctuated the black water like beacons. This was the annual Fourth of July "Jade Green Octopus Hunt."

Seeking their specimens, the students fanned out, picking their way carefully around the ominous spiny sea urchins. Among the rocks and coral chunks were dark grottos, havens of the elusive jade green octopus, *Octopus vulgaris*. "Look for a hole surrounded by seashells and you will find the lair of the octopus, a notoriously sloppy housekeeper," I said. The students staggered off, tripping over rocks and chunks of dead coral, scattering in a tight circle about forty feet across, afraid to separate themselves too far from their comrades. Occasionally someone's light revealed a damselfish

9

protecting its territory or a small green moray sinuously wending its way from shadow to shadow.

A scream pierced the darkness. "There it is!" The rest of the class stumbled across the flats and there, in the glare of a dozen flashlights, crouched a very shocked octopus, its green surface glowing balefully. After a moment it slithered towards the haven of an overhanging rock. Pandemonium broke loose. "Get it!" was repeated by a dozen voices, followed by a veritable stampede of stumbling students. But the octopus had its chance and it disappeared.

Disappointed, the students fanned out again, this time the wiser. Another yell. Another staggering convergence. This time a tentacle was grabbed. It broke off. The co-professor, "the world's largest and funniest marine biologist," had just returned from Japan, where he had eaten a local delicacy, fresh raw marinated octopus. Almost instinctively, he popped the tentacle into his mouth in the customary Japanese fashion. His eyes bugged out and he gagged mightily. The dozens of suckers lining the tentacle attached themselves to his throat in one great reflexive convulsion. He couldn't swallow his delicacy. No matter how much he tried, the tentacle adhered to his throat. The class gathered, gawking at the sight of this huge man roaring and gasping and pointing to his throat. No one knew what to do. Removing an octopus tentacle was beyond the experience of everyone, even those who had taken first aid courses. Finally, with a resounding, grinding roar, he ripped the tentacle from his mouth. Relieved, the class resumed the hunt, leaving their co-professor to slowly wade off toward the distant light of the lab.

Soon the specimen bag bulged with a frantically squirming octopus. It was carried triumphantly back to the lab and placed in an aquarium for later examination. (It soon escaped and was not available for the final exam.)

❧❧❧ Octopussy

A gorgeous woman sits languidly on a throne-like dais. She fades into a blur as the camera focuses on an octopus churning around in its glass prison. The blue rings on its flanks glow as warnings of instant death.

What kind of woman would keep one of the most dangerous animals in the world as a pet? In walks James Bond, and we understand this extravagant panorama of incipient danger. It turns out that Octopussy is benign—not so her pet. The Australian blue-ringed octopus, *Hapalochlaena lunulata*,* has evolved a neurotoxin that paralyzes the phrenic nerve that innervates both the diaphragm and heart. Seconds after the bite, the heartbeat becomes irregular and diaphragm muscles become paralyzed, causing suffocation. Death occurs minutes after injection of the poisonous saliva.

Octopods, members of an ancient class of mollusk, have had millions of years to evolve ever-more efficient poisons with which they paralyze their prey, which is almost exclusively crabs. A large crab can do damage when the octopus attacks it. After a flurry of flailing claws, the crab disappears under the web-like membranes between the tentacles. Then the still-dangerous crab is bitten by the upside-down beak of the octopus (resembling the beak of an eagle, but the lower mandible hooks over the upper). Poisonous saliva flows copiously into the wound. The crab ceases its struggle, instantaneously paralyzed. It is important that the crab stops fighting as soon as possible to prevent injury to the octopus. Consequently, over the eons octopods have evolved more and more complex, more and more effective paralysis-producing poisons.

The Australian blue-ringed octopus has succeeded in developing a venom consisting of ancient and new compounds so complex as to defy synthesis. An antidote cannot be produced in the laboratory. If a human picks up one of these small animals and squeezes it, the alarmed octopus will bite its antagonist. Death inevitably follows within a few minutes of being bitten. Incidentally, this venom is chemically identical to tetrodotoxin, the powerful toxin (1,200 times more toxic than cyanide) in the flesh and liver of the puffer fish (see chapter 13). This is a remarkable example of parallel evolution, where animals of diverse phyla have developed the same complex molecules.†

There is another famous blue-ringed octopus. The four-eyed reef octopus, *Octopus hummelincki*, is so-named because it has two circular blue blotches near each eye. These false eyespots may have survival value as a means of confusing predators. This small species is dangerous only to its

---

* One of four species of blue-ringed octopods, formerly *Octopus maculosus*.

† It has recently been suggested that both species harbor the same bacterium. According to proponents of the theory, the bacterium produce the toxin.

prey, crabs and snails. Its uniqueness derives from the unusual physiology of its life cycle. Upon reaching maturity, the female mates and then retires to a cave to attach her sac-like eggs to the ceiling. She remains in her birthing chamber, blowing oxygenated water over the eggs. She scarcely emerges, only occasionally pouncing on a passing snail and distractedly going through the motions of tearing off the snail's trapdoor-like operculum. Often she does not even eat the exposed flesh of the snail. Eventually she stops leaving her den altogether. After a while, a cloud of tiny newborn octopods flows from the cave. She dies. Her role as procreator has been fulfilled. But what signal tells her that she has finished her role in the circle of life?

Research has unearthed an unusual feature of *O. hummelincki*'s life cycle. A gland in the brain reduces in size during the lifetime of the animal, disappearing near death. It has been hypothesized that this is a "life gland." Its secretion may regulate the length of life of the animal. It was reasoned that if the hormone from this gland could be synthesized, it might be used to prevent death in humans. After a flurry of research this hypothesis was abandoned. But future research in genetic engineering may yet uncover the secret to eternal life from an octopus brain.

The *O. hummelincki* broods her eggs in a cave. In contrast, the female Australian blue-ringed octopus broods 60–100 eggs in a jelly-like mass under her arms for about fifty days, releasing the newly hatched large, mature larvae, called paralarvae, to become part of the plankton. This strategy seems to be a recipe for doom. Producing few offspring and casting them into the planktonic jungle seems to be self-destructive. But these eggs are brooded for a long time and hatch into large, mature larvae—too large to be vulnerable to most planktonic predators. The larvae feed vigorously in the planktonic soup near the surface, becoming widely dispersed. After a few weeks they sink to begin their bottom-dwelling existence. Thus each octopus species survives though using different developmental strategies.

The King of Beasts

The octopus is the king of beasts. Can a lion change color in an instant? Have lions invented jet propulsion eons before man appeared on the earth?

Is the lion the culmination of an evolutionary history of morphological and chemical experimentation that boggles the mind with its variety? Do lions have a "life gland" in their brains that programs them to die? If you were to be bitten by a lion, your chances of survival would be better than after a tiny nip by a blue-ringed octopus.

❧❧❧ The Promised Land

During my quest for photographs of octopods, I had heard of the promised land for all octopus hunters. The Galeta Laboratory of the Smithsonian Institution lies on a drowned ancient Caribbean coral reef, the shallow waters of which cover a formerly exposed limestone shore, pitted and eroded into myriad holes and caves, ideal habitat for octopods. A small, one-story lab lies at the edge of the octopus flats. At that time, the only accommodation was an old mobile home whose walls were covered with cockroaches in a disgusting panorama that reminded me of Ray Milland's d.t.'s in a scene from the movie *The Lost Weekend*.

Octopods reveal themselves at night. We approached the shore, sneaking up on them in the dark. At the last minute we turned on our flashlights. Hordes of octopods scurried away, seeking darkness. I leaped into the water, wildly groping, and caught one. Triumphantly I held it aloft. My wife readied her camera. Then I noticed that the animal was squeezing from between my clenched fingers. I tightened my grip. Still it oozed from between my fingers, leaving a patch of slime on my closed hand. Then, to my horror, I felt its cold tentacles climbing up my arm towards my vulnerable armpit. With a gasp of disgust, I flexed my arm, flinging my specimen back into the water.

After taking a few minutes to recover, I tried again. Again the octopus squeezed through my fingers, threatening my armpit. Again I flung it away. Finally, my all-knowing wife pointed out that I needed gloves. I found a pair and entered the fray again. Soon a nice, juicy octopus was trapped in my gloved hand, its roughness temporarily preventing the animal from escaping. I rushed for the shore, my victim sliming mightily and flailing its tentacles. What to do with it? I peered desperately into the dark night.

Dimly, I saw a series of water-filled tanks along the sides of the building. I rushed to one and plopped the octopus in. Relieved, I turned on my flashlight. To my consternation, the light revealed a sign: EXPERIMENT UNDER WAY. DO NOT DISTURB. There was a screen separating the experimental half of the tank from the control section.

What to do? No choice. Honor demanded that I remove the octopus. After all, this experiment might have been in progress for months. I plunged my hands into the tank. The octopus immediately jet-propelled itself to a corner. I grabbed at it. Chaos ensued. The animal crawled and jetted across the tank, easily eluding my grasp. It climbed up the sides of the tank and hid in corners on the bottom. Flailing about in a blind frenzy, I finally caught it and flung it back into the water.

I looked back at the experimental tank. The screen was ajar. The experimental chamber was violated. Whatever had been separated was together now. It was a mess. We made our apologies and left the next morning, feeling immensely guilty.

Octopods are the most highly evolved members of the phylum Mollusca. They bear little resemblance to their lowly relatives, the clams and snails. A layer of filmy tissue, the mantle, produces the shell in those sedentary classes. It becomes a muscular envelope in the octopus, pumping oxygenated water across the gills, allowing a relatively high rate of metabolism. The ancestral shell is reduced to a few spicules of calcium carbonate embedded in the muscular mantle. The absence of a shell explains the rubbery consistency of the octopus's body and its ability to squeeze through virtually any constriction.

When the octopus is stressed the mantle has another function. Water is taken in when the muscles are relaxed. Folds of tissue close around the edges of the mantle, producing a closed chamber. Then the mantle contracts, forcing the water out through a narrow funnel. The large volume of water from the mantle cavity spurts through the funnel with considerable velocity, jetting the animal backward. Though its usual mode of mobility is to crawl over the bottom, it uses its jet to escape from predators, fan its eggs, and blow away nuisances such as the tiny, irritating, pugnacious damselfish.

The squid, a close relative of the octopus, has an internal "pen," a vestigial shell that has evolved into an internal plastic-like stiffening rod that causes the animal to keep its shape. The octopus has no such restrictions on its shape. Although its bulbous body appears permanent enough, it can assume the shape of any container in which it is placed. Put a six-inch-wide octopus at one end of a two-inch tube and a crab at the other. Soon the octopus will be at the other end of the tube, munching away.

The story is told of a researcher at the University of Miami who was becoming paranoid. This is a common temporary affliction of graduate students, and no one paid any attention. But his complaints were becoming obtrusive and someone listened. He was able to wring this tale from the now blubbering young man: his experiment on crab behavior was being destroyed. Someone was stealing his crabs. Every night an experimental crab was removed from the tank. Who had it out for the poor fellow? It was decided to hide in the lab and watch for the experiment-destroying bad guy.

Late that night the lab was dark except for the glow emanating from lights over dozens of aquaria. Everyone hid behind lab benches. No villain appeared. Wait! An octopus had left its lair at the bottom of an aquarium and was climbing up the walls. It reached the glass cover that had been securely taped to the tank top. There was a small feeding hole in the lid. The octopus crawled to the hole. A tentacle emerged, then another, until all eight were sucked onto the outside of the lid. The body emerged. All that was left inside the tank was the head with its bulbous eyes. After some tugging and with an almost audible pop, each eye flopped through the hole. Then the octopus crawled across the floor and mounted the crab tank. The rest is history.

~~~~ Octopus IQ Tests

Is an octopus as smart as a dog? In the absence of any evidence to the contrary, I am willing to believe this. But an octopus can't bark. We can't ring a bell and wait for the octopus to salivate. How do we find out how smart an invertebrate is? A clam just sits. It eats and burrows. A snail eats and crawls. That's it. Clams and snails are mollusks, and so is an octopus. It evolved from the same ancestor as the sluggish, behaviorally primitive clams and snails.

PLATE 2

A. THE COMMON OCTOPUS, *Octopus vulgaris*, searching for crabs. Usually brown but color varies with the background. Up to 3 feet across including tentacles. A tube-like siphon is visible under the eye. When inhaling, the octopus relaxes the bag-like mantle, a slit-like opening appears on each side, and water enters the mantle cavity and flows over the gills. The mantle wall then contracts, forcing water out of the siphon. This mechanism is also used to rapidly propel the animal backward in time of stress, with the mantle rapidly contracting to produce a strong jet of water.

B. BEAK-LIKE JAW OF AN OCTOPUS. The lower mandible is longer than upper, the reverse of an eagle's beak. This bottom beak hooks under the edge of a crab's carapace (or a snail's operculum) and tears away a fragment, exposing the prey's tissue. Poisonous saliva is then injected.

C. EXPERIMENTAL CHAMBER used to study learning in octopods. The octopus in the right chamber is watching the one in the central chamber figure out how to enter a tube to obtain a reward, a crab. The opening of the tube is high on the wall. With repetition, it takes the octopus less and less time to find its reward. The observing octopus takes even less time than its "teacher" to solve the problem.

D. THE BLUE-RINGED OCTOPUS, *Hapalochlaena lunulata*. Up to 6 inches across including tentacles. Brown or tan with vague rings except when excited—then it turns bright yellow with iridescent blue rings. Its venom can cause death within minutes, but octopus will not attack humans unless touched or squeezed. It can release venom into a tide pool to poison the crabs inhabiting it. Common in Australia, uncommon elsewhere.

E. WHEN AN OCTOPUS identifies the proper combination of paired symbols, it is rewarded with a crab. When it fails, it receives a mild shock. The first three pairs of symbols are quickly associated with a crab. The last pair (diagonal black lines) is more difficult: "three- to four-year-old children, just like octopuses, confuse oblique lines oriented in opposite directions although, like octopuses, they can readily discriminate vertical from horizontal lines."*

* W. D. Russell-Hunter, *A Life of Invertebrates* (New York: Macmillan, 1979), 462.

PLATE 2

A.

B.

C.

D.

E.

Just how smart is an octopus?

Field studies have revealed that an octopus can forage as far as four hundred feet from its burrow and find its way back unerringly *by a different path over a different set of obstacles.*

While feeding on a crab, an octopus can demonstrate three modes of behavior, suggesting that it makes choices. It can apply any of these strategies:

1. Cut a hole in the crab's carapace near a key muscle and inject paralyzing saliva.
2. Rip off the carapace.
3. Cut a hole in the edge of the carapace and inject the saliva.

This sort of variable behavior is at the threshold of thought. All other invertebrates exhibit invariable, genetically determined behavior called taxes. A stimulus evokes the same response, always. A repertoire of responses is unheard of.

A red ball was provided to an octopus. It explored this new thing. It found that it was inedible. Did the octopus ignore this new object like any self-respecting primitive organism? No, it spent inordinate amounts of time manipulating it *as if it was playing.*

Can an octopus think? Can an octopus learn? How can we measure how smart an octopus really is? With limitations, laboratory investigations are useful.

Most learning tests with animals involve a maze. The classic study uses rats and a stimulus of food at the center of a maze. The rat must solve the problem of penetrating the maze to find the food. It succeeds by chance, wandering through the maze and accidentally finding the food. Then the experiment is repeated. The rat finds the food more quickly the second time and increases its speed in solving the maze problem in subsequent trials. We call this learning.

The maze is a series of passages for the rat to run through. How does one make a maze for octopods? One way is to divide a tank in half with a clear plastic wall. A crab is visible through the wall. The octopus tries to attack the crab and is thwarted by the wall. Stimulated, it explores. By accident it finds a hole in the wall opening into a tube. At the end of the curved tube is another crab, not visible to the octopus. It enters the tube by chance and finds the crab. Then the octopus is again placed in the chamber. The test is to see if it purposefully seeks the unseen crab in the tube. The amount of time it takes to find the crab declines after several trials. After three weeks of training one octopus learned to solve the problem within two minutes.

Another experiment places symbols in front of doors. The octopus is faced with five doors, only one of which has a crab behind it. A different symbol is placed in front of each door. After a few tries the octopus accidentally chooses the correct symbol. In subsequent trials, it takes only a few seconds for it to open the correct door to find the crab. If a crab is placed behind a door with a different symbol, the octopus learns the new symbol quickly. Octopods can also recognize the full spectrum of colors (and almost instantly change color to match).

In one amazing study, untrained octopods were allowed to watch trained octopods solve a maze puzzle. When introduced into the maze, *the octopods that had observed the trainees solved the problem faster than other untrained octopods.* Is this an example of learning by watching another animal solve a problem? If so, it is unique in the invertebrate world. Even vertebrates such as fish cannot perform this process.

So far, learning in octopods is "observational learning". The next step is to look for abstract reasoning. Some humans are not quite able to reach this level. Watch out humans, the octopods are coming!

3

Bedtime Stories

ON THE SECOND DAY of our field course the students are faced with a terrible prospect. Some have been known to cry. Their tortured looks focus on a piece of paper tacked to the wall. It is THE LIST. On it are 100 dauntingly unfamiliar scientific names, like *Tripneustes ventricosus* and *Diadema antillarum*. The challenge is to identify each organism and classify it as to phylum, class, order, genus and species—and to describe its ecological niche (role in the environment). The cause of the students' consternation is the offhand comment by the instructor that the information required will appear on the final examination SCHEDULED TO BE ADMINISTERED IN TEN DAYS.

Gradually, fear gives way to resignation. "OK," they say, "give us the specimens." The instructor responds, " I will give you nothing. Look around you. What do you see?" Eventually it dawns upon the students that they will have to search for the organisms in the mysterious sea.

Frenzied activity commences. People run (or swim) to and fro, practically bumping into each other. Eventually order prevails. Someone with leadership capability emerges to organize search teams. (This phenomenon has occurred at virtually the same time and in the same manner for the past twenty years). Specimens begin to appear in the aquaria and on the lab tables. Piles of books are strewn about. Late at night the silence is pierced by students shouting at each other. What are they fighting about? To my delight, they are arguing about whether the animal is an opisthobranch or a

20

prosobranch. This tremendous effort continues until there are only a few nights left before the exam.

It is then that the esteemed instructor enters and the room becomes hushed. In compassionate tones he quietly promises to reward the students with a few moments of respite. It is time to read "bedtime stories." The lights go out, revealing the luminescent deep-sea monsters emblazoned on his T-shirt. In the feeble glow of a flashlight, he reads this story:*

🙖🙖🙖🙖

Three Indian women were sitting in the shallows of a tributary of the Amazon River, languidly washing clothing and gossiping. The water slowly flowed by, its warmth suffusing the atmosphere. The mood was tranquil. Suddenly one of the women shrieked. The other two immediately grabbed her legs and spread them apart. One of them reached into her vulva and ripped out a thin, pencil-like squirming fish. Its gills and fins were extended, necessitating the tearing of the woman's flesh. Blood ran down her thighs. But the pain was unimportant compared to her fate if her friends had not been so alert.

She had been attacked by a species of fish whose detestable life style was the source of rumors that filtered out of the Amazonian jungle of a fish so horrible that it killed by entering the human body through the urethral opening to penetrate into the urinary bladder. There it erects its spines permanently, and so ensconced, begins to tear at the bladder walls until it reaches the blood vessels and becomes engorged with blood. "Once [it] . . . has wriggled up into the urethra, the situation becomes so critical that

---

* This story is a synopsis of a chapter in a book by Edward Ricciuti titled *Killers of the Deep* (New York: Macmillan, 1973). All quotes come from this book. I cannot approximate the sinister style of this excellent author. Much of the original content is included so as to re-create the experience for the reader. For more up-to-date information, consult Stephen Spotte, *Candiru, Life and Legend of the Bloodsucking Catfishes* (Berkeley, CA: Creative Arts Book Co., 2002).

many a male victim has slashed off his penis, preferring life with impaired sexual ability to a painful death."

Hearsay of the existence of this unusual phenomenon began to reach civilization in the nineteenth century. The world of science was incredulous, rejecting the idea of a vertebrate parasite of man. The grisly prospect of a fish tearing up the innards of a *human* host, killing it quickly, went against all theory. The rumor further reported that the host's destruction is so rapid that a cow being driven across a river to market and attacked by several of these parasites became so drained that "within two hours the cow was so weak that it barely could keep its feet, and when its throat was slashed at the slaughterhouse, hardly a dribble of blood remained in its veins." The bizarre stories became so persistent that the common name of the fish, "candiru," was accepted and its fish-to-human interaction became the subject of conjecture in scientific circles. During the early 1900s the rumors intensified. A trained scientist and explorer reported that he "saw Indians in the Amazon don codpieces constructed of plant fibers before entering the water. Other Indian men merely tied the foreskin of the penis over the glans. The tribesmen told the explorer that they took such precautions to protect themselves against the candiru. Indian women who had to enter the water wore a sort of G-string."

The rumors persisted. "A physician in La Plata, Brazil, reported that he had cured three boys and a man. The treatment was instant amputation of the penis." Subsequent investigations by medical researchers reported that Amazonian Indians had a method "that probably was much favored over amputation by males. Medicine made of the juice of the unripe fruit of the jagua tree, *Genippa americana*, caused the candiru to relinquish its hold and slip out of its host."

In 1930 a definitive paper was published, "The Candiru, the Only Vertebrate Predator of Man." The reality of this terrible parasite, *Vandellia cirrhosa*, was established. Examples of candiru behavior trickled into the scientific literature. It was discovered that it could be found in the bodies of dead fish. Apparently it was a carrion-eater, an aquatic vulture. A technique was devised for collecting the candiru. A bloody cow lung was hung in the river. Upon its removal, a hideous mass of candirus squirmed through the bloody mess. "Some . . . bloated with blood and offal, clung to the lung even after it had been removed from the water."

What attracts the candiru to the genital region of the host? Speculation focuses on the urine. It is possible that the fish is attracted to the chemical nature of urine. Fishes excrete urea and ammonia through their gills, and a parasite accustomed to obtaining sustenance from the gills of fishes could mistake a urinating urethra for a gill cavity. Or is it the current created by the flow (fishes swim upstream)?

The controversy was all but resolved by a series of ingenious experiments at the American Museum of Natural History. Four candirus were placed in an aquarium. The experimenter tried a variety of foods, reasoning that if the fish are scavengers they would eat worms, dead goldfish, or even fish blood. They rejected all of these foods. When a live goldfish was placed in the water,

> the candirus swam about searching for it and quickly three of them attached themselves to the gills of the goldfish and began to suck its blood. This was accomplished by the use of long teeth that project from the upper jaw. Burying their heads in the gill chamber, the candirus rasped away through the membranes covering the gills. Flecks of tissue and pale red blood dribbled from the gills as the parasites sucked vigorously. Soon their bodies grew turgid and they dropped off, bloated and motionless on the bottom. Several of the goldfish survived the initial attack, only to expire after participating in further tests.

It is well to remember that these scary freshwater catfishes are found only in the Amazon basin. But this knowledge notwithstanding, there seems to be a reflexive phenomenon that occurs after hearing the story. For days after this bedtime story, the men in my Caribbean class reported that while swimming, the scrotum was reflexively retracted as close to the body as possible.

Is the candiru a predator, as described in the first scientific paper devoted to its study, or a parasite? Vertebrate predators are common, but vertebrate parasites are rare. The vampire bat, invisible in the darkness of night, ends its whispering flight by landing imperceptibly near its sleeping human host. A quick, painless incision with razor-sharp canine teeth, and a red, spreading stain appears on the host's skin. The bat licks up the blood and when gorged, flies off. Its unknowing victim awakens the next morning and feels none the worse for wear, suffering only an inexplicable tiny wound. Vampire bats, despite their notoriety, primarily feed on the blood of cattle, rarely attacking humans.

Most parasitic relationships cause little immediate harm to the host, but virtually all predators are killers. Most parasites are in it for the long term, but most predators destroy immediately. Are candirus parasites or predators? Or neither?

❧❧❧ Living Together

Regardless of the vague boundary between predation and parasitism in this case, there is a variety of unusual relationships that are more easily defined. Sometimes fishes will interact in a mutually beneficial way. Formerly identified by the general term symbiosis, which means "living together," the more precise term is mutualism. A grotesque source of semantic confusion is the male of a species of anglerfish, *Melanocetus johnsoni*, which soon after a brief free-living existence, finds a female and attaches to her underside in close proximity to her reproductive opening. Once ensconced, it is overgrown by its mate's tissues, and becomes one with her, nourished forever by blood vessels that form a plexus around it, just as the human fetus is nourished by a network of blood vessels in the placenta. When it comes time to spawn, the male is always in close proximity to provide sperm. This is of survival value in the almost barren deep sea where this species lives. Finding a mate at the proper moment would be very difficult. The live-in male comes in handy.

Is this an example of mutualism? Is a nursing infant exhibiting mutualism? Or is the relationship one of parasitism? After all, the male anglerfish and the infant are obtaining nutrients from the female. So would a tapeworm buried in her gut. No, mutualism is heterospecific—an interaction between different species. This example, no matter how wonderful, is not considered mutualism because the intimately interacting participants are of the same species.

A valid example of mutualism is cleaning symbiosis (which should be called cleaning mutualism). A small species of tropical marine fish (or shrimp) cleans the host fish, a larger species, of its parasites. The tiny cleaner sets up a cleaning station by disporting itself on a pinnacle of coral and waving to passersby, advertising its services. A large fish approaches. In this complex web of instinctive interactions, the customer assumes an unusual position: head up or tail up or mouth yawning, a patterned behavior signaling that it will not attack. The tiny cleaner picks up the signal and approaches with impunity, swimming first into the gill chamber of the hovering host. It picks off parasitic copepods and flatworms specifically adapted to a life of sucking blood from the vulnerable gills. Then the cleaner exits from the mouth, picking its host's teeth on the way out.

Why is it that the carnivorous host fish becomes a pussycat when it is being cleaned? Does it somehow know the benefits of the interaction? This is not the answer. The neurological apparatus of the fish brain precludes rational thinking and prescience. In the nature vs. nurture battle, score one for nature. Unthinking, the host fish loses its aggressive instincts when it perceives the signals exhibited by the cleaner. It then produces the answering signal, a peculiar stance. The dance continues.

All of this choreography is embedded in the genes. Stimulus A is evokes response A, stimulus B evokes response B, and so on. It is not difficult to envision the evolution of this behavioral ballet. Lost in the mists of time, a defining moment occurred: a small fish, pecking on a rock, came upon another rock-like structure, the flank of a huge fish. Oblivious, it continued its plucking behavior, unaware that this bountiful source of worms and copepods was a predator. In a flash, the inadvertent cleaner became lunch. Over time, host fish that allowed the cleaners to work survived better than the "uncleaned fish." A nonaggression pact was embarked upon. Eventually a ritual evolved, a set of signals of survival value that say "for the time being, peace between us."

The evolutionary process can be studied by visiting coral reefs in the Caribbean (Atlantic) and off Australia (Pacific). The Caribbean fauna has had a comparatively short time to evolve, since ice ages made the water too cold for the survival of the nurturing coral reefs, which need the warm water of tropical seas to exist. The reefs of the South Pacific, by contrast, were so insulated by distance from the icy water of the nearest melting glaciers that the water remained warm and evolution continued unabated. Thus ancient processes prevailed in the Pacific and newly evolved behaviors (many

PLATE 3

A. FEMALE TORSO showing the fate of the Indian woman if not for the alertness of her friends. Once the fish enters the urethra, there is no obstruction to its passage into the urinary bladder. The small fish can be seen inside the bladder surrounded by richly vascularized walls (shaded). The fish tears at the walls and ruptures blood vessels. The fish is able to survive, but it is trapped inside its host until the host dies. No well-adapted parasite kills its host, suggesting that this is a recent association and has not yet evolved into a truly functional parasitic relationship. There is only circumstantial evidence that the candiru can reach the bladder in human males because male anatomy presents obstacles.

B. CANDIRU, *Vandellia* species. About 2½ inches long and ¼ inch wide, dark brown to tan. Upper jaw contains long, recurved teeth capable of tearing tissue.

C. DORSAL VIEW OF HEAD OF THE CANDIRU, *Vandellia hasmanni*. The small spots are nostrils. The larger ones behind them are the small eyes. Candirus have poor vision. Their sense of smell is the primary mechanism for finding food. Small projections along the edge of the head are barbels, similar to those of other catfish, used to identify food by touch and taste when in a scavenger mode. They project from the lower jaw next to the mouth. Two pads of spines can be seen at the back of the head. They face backward and make the candiru difficult to remove. Another pair of spiny pads is located on the underside.

D. A CANDIRU feeding on goldfish placed in an aquarium. It circles its prey, touching it, then enters the gill chamber and hangs from it. Eventually the bloated parasite drops from the goldfish to digest its blood and tissue meal.

E. CLEANING WRASSES removing parasites from the gills and teeth of a large grouper, an example of mutualism. When the grouper presents the proper signal, the cleaners will enter its mouth or gill chambers and eat harmful parasites. Cleaners get food and the grouper gets cleaned.

PLATE 3

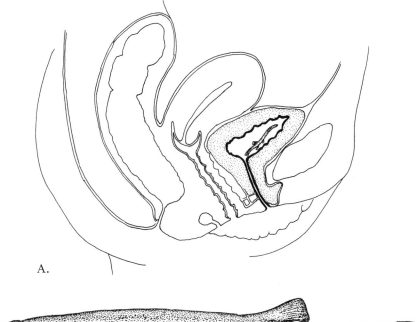

A.

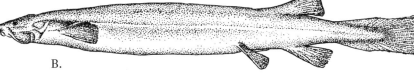

B.

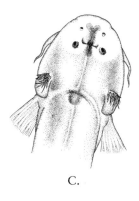

C.

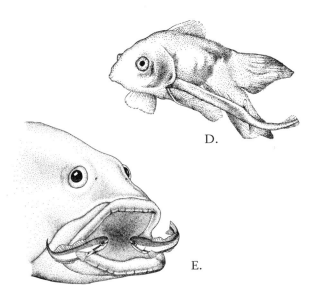

D.

E.

in the 10,000 years since the last glaciation) arose in the Caribbean (Atlantic).

Caribbean cleaners don't need to clean fish to survive. They usually behave in the traditional manner, that is, plucking small organisms from the
surface of rocks. Occasionally—no one knows what the stimulus is—they
offer their services as cleaners. They set up a cleaning station on the tip of
a coral head with a coterie of customers waiting in a long line. It is not uncommon for divers to happen upon what appears to be a parade of various
species of large fishes, each hanging in the water and assuming the peculiar
position of passivity to invite cleaning—customers for the fishy barber.
This is in contrast to cleaning mutualism in the Pacific. There, the cleaning
wrasses *Labroides* spp. *must* obtain their total sustenance from cleaning.
They are obligate mutualists and the Caribbean cleaners are facultative
mutualists.

In fact, the intimacy between the Pacific cleaners and the cleaned is so
profound that there is a marked decline in populations of large fishes when
the cleaners are removed from a section of reef, and those remaining have
ragged fins and exhibit other symptoms of unhealthiness. In turn, the decline in host fish numbers leads to a further decline in cleaners, leading to a
death spiral for both.

4

Garden of Eden: The Death Apple and the Tree of Life

LONG AGO AND FAR AWAY there existed a tiny island where cooling trade winds always blew, yet the sea was glassy calm. A little "mountain" a few hundred feet high suggested aspirations to join the other British Virgin Islands. It almost succeeded, but for a watery barrier between the island and the rocky shores of Virgin Gorda. A lazy stream of seawater flowed perennially through this channel, so narrow that you could almost shout across it. A current evoked by the wind's long fetch down the broad Sir Francis Drake Channel met tidal waters pouring from the surrounding embayment, the confluence carrying planktonic riches to sustain coral growth so magnificent as to create a splendid underwater scene. Six-foot-long tarpon, their iridescent scales glistening like a treasure trove of silver coins, crowned the watery vista. They elegantly circled unbelievably dense clouds of anchovies and silversides.

On the other side of the island, a twenty- minute walk over an elegant stone walkway to nowhere, lay a rocky path. Carefully picking your way along the treacherous trail, you found yourself on a promontory high above the sea. Your eyes, focused on picking out safe footholds among the scrabble, opened to an incredible sight: a tiny pristine beach no more than fifty feet across. The sand was whiter than any I had encountered elsewhere, the sea more azure. Looking down one could see a great variety of fishes richly embellishing this aquatic paradise. It was called Honeymoon Beach. If the easterlies swung out of the northern quadrant, it was tranquil here; if out of the southern quadrant, it was smooth at the channel between the islands. One could choose a place of tranquillity every day. Top this off with a hotel consisting of four or five classy cabins with a cordon-bleu chef and a seaside bar. I happened upon this secret place while sailing along the Sir Francis Drake Channel in the British Virgin Islands. It was to be the source of my inspiration for the next forty years.

❧❧❧ Lesson One: How to Defend Yourself against Rapists

I vowed to return. I did, sharing my paradise with a generation of students. News of the course spread. Soon nonstudents joined us. One of these was a spiritual woman in her forties. She had come to the island with us to find inspiration to make a decision: whether or not to expand her mail-order business into a retail outlet. She offered complete independence for women who were tired of the tribulations of male-dominated relationships. Her version of freedom consisted of a variety of high-tech devices that were guaranteed to deliver sexual emancipation from men. The shop was called something like "Venus's Girdle." She was ahead of her time, riding on the cusp of the wave of the women's movement, and she questioned whether or not there was enough clientele for her wares.

She had little interest in my lectures and spent her time in solitude seeking the resolution to her problem, wandering through the beauty of the island. One day she returned from one of her walks and joined me at lunch. She picked at her food, seemingly lacking an appetite. I inquired, and she told me this:

"I was walking along a path and I came upon a tree with sweet-smelling apples. I plucked one (visions of the Garden of Eden had crossed her mind) and ate it." I almost fell off my chair in consternation. She had eaten from the manchineel tree, *Hippomane mancinella*, whose sweet smelling fruit is called the "death apple." The tree is so toxic that those who take shelter from the rain under it develop severe rashes from the poison in its sap. Remembering the death of a member of a Danish Olympic team on a nearby island who had eaten several apples, I gasped, "How many did you eat?" She said, "Just one, and it didn't taste good."

It turns out that she had taken a course on how to repel a rapist, and had learned the most effective method: vomit on your assailant and he will leave you alone. Using her liberation lore, she had vomited out the death apple. I never found out what her decision was, but her decisive actions in paradise saved her life.

The manchineel apple has evolved a survival mechanism different from thorns or a bad smell. It produces a toxin so powerful that an apple, once tasted, repels goat, pig, or crab (either by death or an induced aversion). The poison from the tree's caustic sap causes skin lesions and destroys mucous membranes if eaten. It was used by aborigines as an arrow poison. The native antidote is arrowroot. In The Grenadines, a poultice of arrowroot in water is applied to the burn-like rash. This small tree is found in coastal thickets and mangrove margins throughout the Caribbean.

ﺷﻮﺷﻮﺷﻮﺷﻮ Land Reclamation Project

> Like the many-armed Hindu god Shiva,
> the Red Mangrove tree squats in the mud
> and creates land out of the void. Its many
> prop roots reach out into the water, and
> tiny motes of sediment fall at its feet.
> The mystical belief in the cyclical nature
> of creation and destruction is also fulfilled,
> for the very nature of its creative role
> spells the doom of this unusual tree.*
> —EUGENE H. KAPLAN

The deadly manchineel tree notwithstanding, the shoreside trees of the Caribbean are benign. The spidery red mangrove, *Rhizophora mangle*, is one

---

*Eugene H. Kaplan, *A Field Guide to Southeastern and Caribbean Seashores* (Boston: Houghton Mifflin, 1988), 173.

PLATE 4

A. A 30-FOOT HIGH EGRET TREE, A RED MANGROVE, *Rhizophora mangle*, provides shelter for hundreds of cattle egrets. In this scene the birds have risen en masse from the tree, as if by some unknown signal. This often happens several times before they settle down for the night. Prop roots and aerial roots can be seen in the intertidal zone.

B. PROP ROOTS OF A RED MANGROVE TREE. Barnacles and oysters demark the intertidal zone. The subtidal zone, with its anemones and sponges, is perpetually under water. The camouflaged crab is a mangrove tree crab, *Aratus pisonii*. The animal with two chimneys is a tunicate. The snail is a mangrove periwinkle, *Littorina angulifera*. The encrusting sponge with large pores is a bright red or orange fire sponge, *Tedania ignis*, so called because it causes a burning rash. The dark-red alga festooned on roots is a spiny seaweed, *Acanthophora spicifera*. A sea anemone and a fanworm are visible in the center.

C. RED MANGROVES COLONIZING THE SHALLOWS. Seedlings may float in the sea for a year before becoming grounded on a sandy shoal. The pointed bottom becomes embedded in sediment. Small roots and leaves sprout, then prop roots that impede water flow, causing the sediment level to rise until it is dry land. Most seedlings sprout near the parent tree. Seedlings on the left have become small trees.

D. SEEDLING, LEAVES, AND FLOWERS OF THE RED MANGROVE TREE. Leaves are shiny, dark green, rhododendron-like, to 3 inches long. Flowers are yellow and waxy, in groups of four. Seedlings, called sea pencils, grow to 10 inches and have a waxy upper region (including brown fruit) and a spongy, pointed bottom. The top repels water and the bottom absorbs it, causing the seedling to float bottom down.

E. MANCHINEEL APPLE, *Hippomane mancinella*. Fruit to 1 inch in diameter, light green or yellow, sometimes with reddish blush. Ripe fruit smells sweet. Tree to 30 feet tall, has shiny dark-green leaves and inconspicuous flowers. Fruit and milky sap are very toxic. Taking shelter under the tree when it is raining results in a painful rash caused by sap washing from leaves.

PLATE 4

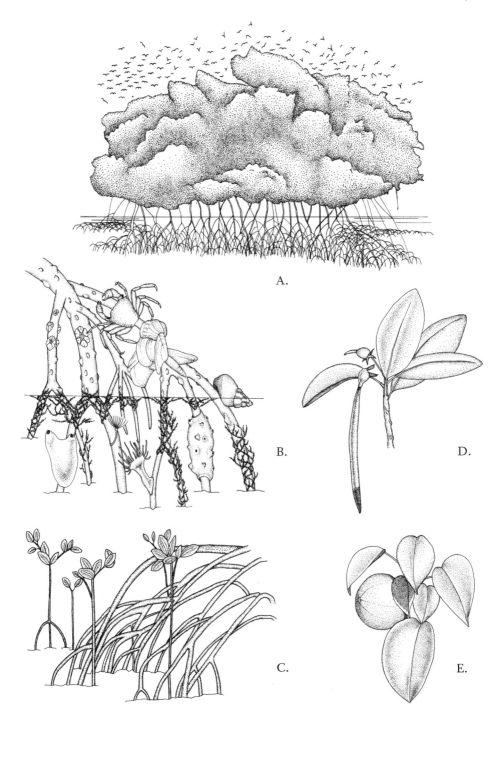

A.

B.

D.

C.

E.

of the most important of the earth's trees. With a huge rotund mass of rhododendron-like leaves, its forests become dense morasses of twisted waterways. Its lifestyle echoes the survival mechanism of the brine shrimp. How do tree and brine shrimp resemble one another? By virtue of their vulnerability, tree and brine shrimp can live only in environments so inhospitable that their competitors and predators cannot survive. The brine shrimp lives in water too salty for other animals. The mangrove tree lives in full-strength seawater, something that other trees cannot do. In each case the salty water creates what is called an osmotic disequilibrium that they deal with by either pumping out the salt or preventing the salt from entering.

The mangrove is a terrestrial plant that must bathe its roots in the sea. It can live on dry land, but in the end it will be outcompeted by other trees and its population will disappear. Like any pioneer species, it is short lived, and if typical tropical hardwoods with the longevity of the climax forest intrude themselves among the mangroves, eventually no mangroves will be left.

For its survival, the red mangrove needs, literally, an edge. It lives at the seaward margin of tropical islands. It survives by preventing salt from entering its root surfaces, thus maintaining normal salinity for cellular functions while living in inhospitable levels of salty water.

But another danger looms. Hurricanes are common in the tropics, threatening to devastate the seaside forest. Evolution has provided. Masses of arching, skinny "arms" spring from the trunk and embed themselves in the mud. These prop roots are joined by serried ranks of aerial roots, descending from the tree's branches. This tangled mass of partly submerged spear-like roots forms an impenetrable thicket, preventing the tree from being uprooted by all but the most destructive storms.

These sturdy roots have a multiplicity of functions other than protection from storms. They project from the sea's surface to the bottom, forming a rickety intertidal and subtidal zone. Most animals that fall, crawl, or undulate into the soft sediment beneath the roots will suffocate in the oxygen-poor silt. But just above, in the oxygenated water, a wonderful array of animals and plants has found a safe haven on the mangrove roots. Oysters, crabs, sponges, seaweeds, tunicates, and worms colorfully colonize the prop and aerial roots, forming a fuzzy mass enshrouding the roots and presenting an impediment to the flowing tides.

This "mother of all trees" literally creates land where there was none before. Current carries sediment—the slower the current, the smaller the particle size that can be carried. The organism-covered mangrove roots present a blurry barrier to the current that slows the water down to a scarcely perceptible flow. Since the water is made to move very slowly, the tiniest particles fall like microscopic snow to the bottom, eventually to become a thick layer of soft, oozy mud so velvety that it feels good when you sink to your knees while wading among the roots. The tree is literally creating land by causing the particles of sediment to drop from the water column. Eventually the edge of the land extends outward, and so do the trees. Land reclamation can extend the shores of tropical islands a hundred feet per century. Columbus's ships, scuttled along the shores of St. Ann's Bay in Jamaica in the fifteenth century, are entombed under the mud of the mangrove forests that extend seaward a quarter mile beyond their burial site.

The mangrove tree has evolved yet another miracle. Its waxy flower turns into a fruit that remains on the tree to germinate into a seedling. *The tree broods its young*, protecting the seedlings by suspending them from its branches over the sea. It is as if an apple tree would retain its apples until they turn into miniature apple trees.

After months of suspension, the seedling falls from the branches into the sea. The fruit is a brown knob at one end of the seedling. The other end of the elongated green "sea pencil" eventually becomes spongy and absorbs water. The waxy fruit end repels water; the spongy tip soaks it in and becomes heavy. The waterlogged pointy bottom sinks. The seedling floats vertically, pointed end down, until it encounters a shallow area, sometimes miles away from the parent tree, usually only a few feet away. A wave drives the pointed end into the sediment, and a new tree is born. Roots appear, waxy rhododendron-like leaves project from branches, and photosynthesis commences. Other seedlings dig in nearby. Prop roots from these juvenile trees reduce current flow, depositing sediment. Soon new shoals, then islands appear. These coalesce. The shore extends.

Over the years the land builds up. It rises above tide level. Other trees take root and outcompete the mangroves. A shoreside scrub forest takes the place of the pioneering mangroves. The cycle is complete.

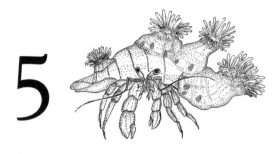

5

A True Romance Story

THE SHIMMERING GLOW of the full moon created a dappled path for me to follow. Pinpoints of light off in the distance were my destination. The night was soft and warm. The young lady in the bow of the boat smiled coyly at me. I blushed. She had won my heart that day by fearlessly wallowing in the mud flats up to her thighs. I had hoped that once her mud-flecked face was washed, beauty would be revealed. It was a relief to see that she had cleaned up nicely. We were destined to dance the night away at the only bar in the town of Friday Harbor. It was approachable only by small boat, and I had requisitioned an outboard skiff for the occasion.

Later, we returned to the boat. Thoroughly drunk, I teetered on the dock holding my hand out in a gentlemanly fashion. She accepted and delicately leapt into the bow. We were on our way back, images of seduction in my mind—when the engine cut out. I yanked on the starting cord until I was exhausted. Visions of my damaged image edged out the picture of seduction. Ignominiously, I began to row back to the lab. A frown was visible on the maiden's brow. An hour later, I pulled up to the dock, exhausted, almost relieved as my date flounced out of the boat and walked off.

The rule for boat usage was to remove the outboard motor and place it on the dock. Raging over lost love, I unscrewed the motor from the transom. It was heavy. The boat was loosely tied, necessitating throwing the motor onto the pier with force—too much force in this case. It landed on the dock—and bounced off into the black, freezing-cold water.

What to do? Honor required that I recover the motor—but how? I stood there scratching my head. It was near midnight. Somebody saw me staring dejectedly at the water. He smiled. His look was puzzling. Was he my savior—or some sort of sadist? It turned out that he knew someone who could help me. The man would loan me some sort of new-fangled gadget called scuba. (It was summer, 1954. Jacques Cousteau had invented scuba a mere eleven years before.)

The rumor of my mishap spread. Someone appeared with a black rubber suit and a metal tank to which were attached two hoses and a circular device with a rubber mouthpiece. He helped me put on the apparatus and departed, leaving instructions to wash his stuff with fresh water when I was finished. I turned the valve and drew a tortured breath from the primitive regulator. By this time something of a crowd had gathered for this impromptu demonstration of the near-miraculous apparatus that allowed one to breathe underwater. Someone put a rope in one of my hands and a huge, very bright underwater light in the other.

Thus equipped, I descended into the icy water. There, in the glow of the light, lay the motor, fifteen feet below. A thrust of the flippers brought me to the bottom. I began tying the rope. I had just about completed the first part of a half hitch when, newly confident, I took my eyes off the task to look around. The black waters were illuminated by my light as if by day. Surrounding me was a seascape of giant furry-topped "flowers," each the size of a gallon milk container. They reflected the light, some white, some bright orange. I recognized the "flowers" as the sea anemone *Metridium*. I had never seen these animals alive—undergraduates study them as pickled, smelly lumps of flesh.

The scene was overwhelming. As far as the boundary of the penumbra of the intense light, the bottom was dotted with dozens of exotic animals. I was amazed. I opened my mouth in awe—and it flooded with water. I realized I was drowning. What should I do? I had received no instructions. Abandoning the rope, I thrust toward the surface, piercing it like a breaching whale. Everyone began cheering, seeing nothing apparently wrong. They seemed to assume that strangled gasping was the norm when surfacing from a scuba dive.

I gingerly pulled on the rope. The half of a half hitch held, and I pulled up the motor, much to the approval of the crowd. The motor was saved! I saw my date peering from the edge of the crowd. I coughed and gasped one last time and stood up, striking a heroic pose. My image was restored!

ꝃꝃꝃ Deadly Flower Gardens

Sea anemones are an anachronism. Hundreds of millions of years ago, predatory animals had not yet appeared. Without predators with anterior eyes and mouths to chase them, the oceans abounded with newly evolved protolarvae and ancient protozoa. Predatory behavior is characterized by the purposeful seeking of prey. Predators with "fronts" would eventually decimate the overwhelming masses of minute plankton, but not yet.

The oceans were a thick soup of microscopic animals. It was not necessary to have eyes and a rapacious mouth to catch prey in this environment. The dominant carnivores had evolved another method of extracting nourishment from the water. They were radially symmetrical—pie shaped—with appendages projecting outward in every direction, ready to extract the abundant prey from the soupy currents.

Sea anemones extended dozens of tentacles, each armed for its full length with microscopic poison darts, nematocysts. Tiny animals spiraled through the water, brushing the tentacles to trigger explosive cells from which burst coiled, thread-like nematocysts. Some of these threads were not toxic. They were gluey with mucus and adhered to the prey, or whiplike, wrapping around exoskeletal protuberances. Periodically, the food-laden tentacle was wiped across the mouth, located on the surface of a central disk. If the swallowed prey had any parts that were indigestible, this was a problem. Sea anemones, to this day, have not evolved an anus. Only one solution: undigested fragments of food pass out of the mouth.

The sea anemone is a throwback to ancient, seething seas. How is it that anemones have survived over the eons, to prosper in virtually every marine habitat, including the almost sterile tropical seas? It seems that the ancient invention of nematocysts has conferred this survivability. The nematocysts have been elaborated over time into more than twenty types, but the three universal mechanisms—poison darts, sticky threads, and whip-like entanglers—still dominate.

✻✻✻ The Marine Mafia

Sea anemones are cnidarians. They possess the phylum's weapons. The big guns of the cnidarians are for hire. The cnidarian "mob's" protection racket offers the use of the nematocysts to a variety of species.

Sea anemones, thought to be pillar-like unmoving sentinals of the seas, can get around. My pet anemone Julius wanders all over the aquarium on a daily basis. But a faster way to travel is to hitch a ride. The fuzzy tricolor anemone, *Calliactis tricolor*, attaches itself to a passing star-eyed hermit crab, *Dardanus venosus*. There is a key stimulus required to initiate the relationship. The crab must tap the anemone, an action that causes the anemone to crawl onto the back of the crab. This is a tiny community of animals banded together for survival: the crab, the anemone and the (dead) snail shell in which the crab lives. Not surprisingly, the crabs will fight each other to steal the anemones, for it has been shown that they protect the crabs from octopods. In the ultimate crab-anemone protective interaction, one crab species plucks anemones from the bottom and attaches them to its claws. When attacked, it thrusts its anemone-armed claws at the predator. What do the nematocyst-armed anemones get from the deal? They are carried around and enter new areas rapidly, and the crab provides a banquet from its table in the form of discards left over from its meal.

✻✻✻ Fidelity and Infidelity among the Tentacles

Many animals take shelter under the nematocyst-covered, always dangerous crown of tentacles of sea anemones. Pairs of the red snapping shrimp, *Alpheus armatus*, live under the canopy of tentacles of the ringed anemone,

Bartholomea annulata. When the anemone is poked, they rush from their haven in its defense.

In an ingenious study, an undersea investigator, working at depths around sixty feet, examined the fidelity of the shrimps by marking them with dots of india ink under the carapace. She poked the anemones, inducing the shrimps to come out, then injected the same number of dots into "couples" under a labeled anemone. Returning the next month, she found three-spot philandering with two-spot under his/her anemone. Rarely did a pair remain together for more than a month.

Some animals benefit from exchanging partners because it increases genetic variability, the stuff of evolution. But mate-swapping has its disadvantages. The interaction is so ephemeral that bonding is not strong. The price of infidelity is lack of protection.

The other extreme is the lifelong relationship between the diminutive, fierce male clownfish, *Amphiprion*, and its larger bovine mate. The aggressive male will defend the female and nest to the death. But this is usually unnecessary. Defense is taken up by the sea anemone. The clownfish hides in the tentacular ring of the anemone, using a protective coat of slime to escape the nematocyst toxin. A pair lives in one anemone in a lifelong relationship. If the female dies, the male becomes a female and a new male joins her. They rarely leave the confines of their territory, the anemone. Eggs are laid in the safe shadow of the tentacles. Periodically, the fishes rise up from their anemone lair to hover over it and pluck small organisms from the water.

But what does the mafiosa anemone get out of it? A preposterous idea is that the clownfishes, suspended over the anemone, lure large predators into the deadly tentacles. But any fish large enough to eat a clownfish would easily escape from the anemone's relatively weak nematocysts. Besides, it would be too large to pass through the anemone's mouth. In fact, by trial and error predators would learn to avoid the anemone and its fish. No, this hypothesis is unacceptable.

What other explanation is there for this relationship? We found the answer by accident. In one of my classes I required that each student produce a one-minute video of invertebrate behavior. One of the students bought an anemone and its clownfish pair from a local aquarium store. He fed the fishes large chunks of goldfish that he expected them to eat. Ignorance, as sometimes occurs in science, was rewarded. The huge pieces of food were not accepted by the clownfishes. They have tiny mouths adapted to eating plankton. As he watched, to his surprise the clownfishes picked up the pieces of goldfish and dropped them onto the tentacles of the anemone. *They were feeding the anemone*! Since that time, this interaction has been reported in several publications. The clownfish's genes dictate this anemone-feeding interaction, just as they dictate its repeated penetration of the tentacles to induce it to produce its dense mucus protective sheath. The deadly nematocysts protect the clownfish, but it pays for its protection with chunks of food.

✎✎✎✎ The Ctenophore Conundrum

1954. Friday Harbor Marine Laboratory on an island in Puget Sound off the coast of Washington state—the same time and place as the outboard motor story. Thrilled at being able to study at this illustrious laboratory, I was, nevertheless, lonely. Standing isolated at dusk on the pier, I reflected wistfully on how far I was from home. Suddenly, my chest glowed with iridescent green goo from which emanated a fishy stench. I looked up. Two fellow students were flinging four-inch, glowing, football-shaped blobs of jelly at me, waving and laughing. I smiled as I wiped the brightly shining slime from my chest. I had been accepted!

The goo was the gelatinous body of an animal that had climbed the evolutionary ladder one rung above the sea anemone phylum, Cnidaria. The jelly-like body has not been discarded, nor has the radial symmetry, but this phylum has evolved a "front," freeing itself from a sedentary life and suggesting the purposiveness of actively chasing prey. In this ancient phylum, Ctenophora, the evolution of an anterior end is just a suggestion of its potential. The animal has a crude sensory structure, a proto-brain that controls the movement and orientation of the body. Strangely, it is found at the aboral (posterior) end, revealing that although the animal swims mouth forward, there is no accumulation of sensory structures in the front charac-

PLATE 5

A. A STARTLING MIDNIGHT UNDERWATER SCENE showing white and orange frilled sea anemones, *Metridium senile*, among sea cucumbers and sea stars. Some of these anemones can be 18 inches tall and round as a bucket. The squat shapes in the background are typical. The "frill" is composed of about a thousand tentacles.

B. TRICOLOR ANEMONE, *Calliactis tricolor*, on the back of a star-eyed hermit crab, *Dardanus venosus*. In this mutualistic relationship the anemone is transported to new habitats and eats scraps from the crab's meals. The crab is protected by the anemone's nematocysts. These are so powerful as to drive away octopods. The crab is 1 inch long and has blue eyes with star-shaped pupils. The anemone has a thick fringe of fuzz-like tan or white tentacles. The column of the anemone has rows of maroon warts, which the crab taps to signal the anemone to climb onto its shell.

C. CLOWNFISH, *Amphiprion* species, among tentacles of a protective sea anemone. Clownfish to 4 inches long, orange with white stripes. The anemone becomes the territory of the clownfish, which lives out its life among the tentacles, laying its eggs under this protective canopy. The clownfish, in turn, feeds the anemone scraps of food too large for it to eat.

D. RED SNAPPING SHRIMP, *Alpheus armatus*, under the tentacles of a ringed anemone, *Bartholomea annulata*. Pairs of shrimps live in a mutualistic relationship with an anemone, coming out to protect their host when it is threatened. They, in turn, are protected by the anemone's tentacles, which are ringed with white bands of nematocysts. The tentacles are brownish, translucent, to 5 inches long. The shrimp is red, to 2 inches long, with three white spots on carapace. One claw is swollen, identifying it as a snapping shrimp.

E. CTENOPHORES, or comb jellies, have eight meridional rows of bioluminescent ctenes (comblike rows of cilia) that light up brightly at night if the animal is disturbed. Sea gooseberry, *Pleurobrachia pileus* (right), to 1 inch long with two tentacles. Beroe's comb jelly, *Beroe ovata* (left), to 4 inches long, lacks tentacles. Ctenophores do not sting.

PLATE 5

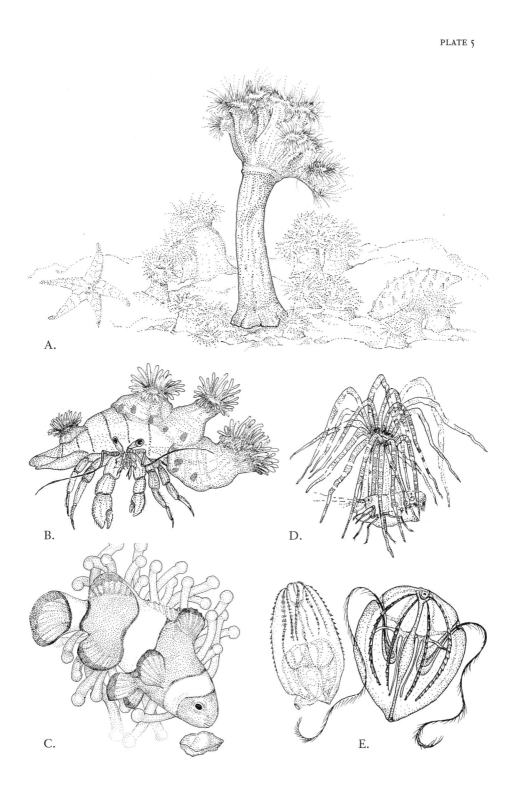

A.

B.

D.

C.

E.

teristic of the more advanced phyla.* As a consequence, the phylum has not yet achieved the ability to chase prey. By swimming randomly through the planktonic masses, huge mouth widely agape, the carnivorous ctenophore is able to efficiently gather in its dot-like food. It still has not evolved a brain or circulatory system, but it seems to have evolved *two* anuses.

Ctenophores swim by means of vestiges of those ancient propulsive structures, cilia, and resemble huge paramecia. But these cilia are organized into rows and radiating meridians. Each row resembles a comb, hence the name of the phylum (Greek *ctene* = comb) and its common name, comb jelly. The microscopic combs are arranged one after the other in circular rows that encompass the body in eight meridians—always eight. This re-markable animal lacks a real brain, yet the beats of its combs are coordi-nated like the rowers of an ancient Viking warship. The rows of combs, beating in unison, propel the animals in what appear to be straight trajecto-ries. But these are misleading, for they are blind, blundering random move-ments. As all plankton, they are at the mercy of winds and currents.

During the day gleaming iridescent rows of combs turn the ctenophores into dazzling jewels. But it is at night that they come into their glory. When rowing a boat in a ctenophore-filled sea, each stroke of the oar produces a stream of tumbling incandescent lightbulbs. The "filaments" are the eight bioluminescent meridians, each glowing brightly as the disturbed ctenophore tumbles backward in the wake of the boat. Why would the ani-mals produce their cold light only when disturbed? If it is to warn off ene-mies, it is too late to light up when passing down the throat of the predator, so the light is not a warning. In fact, if a sea turtle were to be munching on a ctenophore, its activities would disturb the water and cause nearby di-aphanous ctenophores to become luminous and lose their near-invisibility. It cannot be to attract a mate: these animals reproduce by casting almost in-finite numbers of gametes into the water to meet by accident. Remarkably, although it produces light, it cannot see. Ctenophores use primitive light receptor cells on their surface to differentiate light from dark. The vague

* My mentor Benjamin Coonfield described bristle-like touch sensory cells accumulated on the "lips" of the anterior mouth, suggesting the precursor to the accumulation of sensory structures at the front.

glow of the daytime sky attracts and dark repels. Why do they produce their cold light at all? I don't know. Can you figure out this conundrum?

Ctenophores tend to fill the waters with their gelatinous presence at certain times of the year. They do not move into a region as do schools of fishes migrating up the coast, absent one day and teeming the next. No, the process is distinctly an invertebrate phenomenon. Uncountable millions of ctenophores tend to mature at the same time, arising from innocuous gelatinous bumps on the bottom. The presence of huge numbers of organisms as if by a signal is called a bloom, and the human swimmer is often engulfed in a mass of jelly-spheres when ctenophores bloom. Not to fear, ctenophores do not possess the stinging nematocysts of the jellyfish.

My introduction to phylum Ctenophora in the flesh had been abrupt. Gelatinous footballs went whizzing by my head like flying Jello. My soon-to-be friends had plenty of ammunition. There was a bloom of ctenophores.

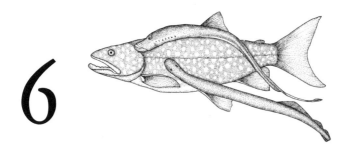

6

Elixir of Love

THE YOUNG MAN knocks on the door, a bouquet of sweet-smelling flowers in his hand. The young lady on the other side of the door dabs a drop of scent behind her ears. She coyly accepts his gift. "They smell so lovely," she says. He replies, "Not half so lovely as you smell." Smell is a strong sexual instrument. The flowers and perfume are designed to stimulate sexual desire through subtle olfactory suggestion. Viewed abstractly, this is not so subtle. The young man is handing his beloved *the sexual organ of a plant*, the flower. She is using a concentrate of the substance used to attract insects to the flower in order that they may transfer the sperm of the flower (pollen) to the egg (ovule).

A fish, a crab, a worm, and a human have one thing in common, the ability to produce a scent so fleeting and ephemeral that it can be detected with special olfactory receptors only by members of its species. A few molecules floating through the air or water can elicit a flurry of activity, usually initiating pre-copulatory behavior. This substance is called a pheromone.*

---

* It is commonly thought that pheromones are species specific—that their effect is felt only by members of the same species—but what if that isn't true?

ᵉᵞᵉᵞᵉᵞ Subtle Sexual Stimulants

Instinctively, humans employ pheromones in order to initiate pre-copulatory behavior. He or she dabs a little mixture of flowery aroma in areas where body heat will maximally evaporate the perfume, releasing it into the atmosphere to attract members of the opposite sex. This substance emits a floral scent, but buried deep in the sweet smell are a few molecules of a pheromone, all that are needed to act as a subtle sexual stimulant. But little does the would-be seductress or seducer know of the source of the potency of the *Evening of Passion* or *Ecstasy* dabbed behind the ears. There is an attractant buried deep in human sweat and emitted from under the arms and the pubic region, but today's highly sanitized human body has lost its olfactory sexual stimuli. Humans must rely on the secretions from the pubic region of other animals masquerading under the sweet smell of perfume.

Male lions, tigers, dogs, and cats mark off the outermost boundary of their territory by urinating on a tree (or fire hydrant). Humans are also territorial. Sexual dimorphism, in which the male is larger than the female, is a sure sign of territoriality. Whether or not humans ever urinated to denote the corners of their territory will forever be lost in the obscurity of time.

Studies of women living in dormitories are instructive. They reveal that, in time, a congruence of menstrual cycles occurs. The timing of the cycles would normally happen at random, but the onset of periods gradually becomes closer and closer among women living in close proximity. What invisible, unnoticed substance would cause this coordination? Possibly pheromones.

Sexual coordination abounds in nature. A school of mackerel has migrated for months and has covered thousands of miles. During this time gonads have been maturing. All at once a female releases her eggs. As if by a signal other females release eggs, and the males release their sperm. Fertilization occurs. If the females had not released their eggs en masse and the males not released their sperm at the same time, the likelihood of sperm meeting egg would be minimal. Sexual products would be scattered all over the sea. Eggs would go unfertilized and sperm would be wasted without

PLATE 6

A. SEA LAMPREY, *Petromyzon marinus*, attached to the flank of a lake trout, *Salvelinus namaycush*. The lamprey probably finds prey by sight. The oral disk keeps the parasite attached to its host no matter how hard it struggles. Trout are often found with healed scars signifying recovery from a lamprey attack. Lampreys are not host specific, parasitizing a variety of fishes, especially lake trout. Three pounds of blood are needed to feed a sea lamprey from larva to adult. It has been estimated that a landlocked sea lamprey kills an average of 18.5 pounds of fish in its lifetime.

B. SEA LAMPREY, *Petromyzon marinus*. Gray, to 14 inches long. It remains attached to its host and feeds for an average of seventy-six hours, dropping off before the host dies and leaving a deep wound that may become infected. Weakened trout die, while strong specimens live on with scarred flanks. Lampreys may have been indigenous to Lake Ontario, remaining in balance with fish populations, but when they entered Lake Erie, they spread to the other Great Lakes, decimating lake trout populations. They normally live in oceans, migrating up rivers to spawn, like salmon. When they became adapted to living in the Great Lakes in the twentieth century, their numbers burgeoned until commercial fisheries failed.

C. MATING FROGS, showing proximity of sexual openings and black-and-white egg mass. This posture, where the male embraces the female, is called amplexus. Pheromones probably attract the male to the female and stimulate release of sperm and eggs.

D. THE MOUTH OF THE SEA LAMPREY is located at base of a cone, the buccal funnel. The buccal funnel and a protrusible "tongue" are studded with sharp teeth. After the host's skin is abraded, an oral gland secretes an anticoagulant, preventing the host's blood from clotting.

PLATE 6

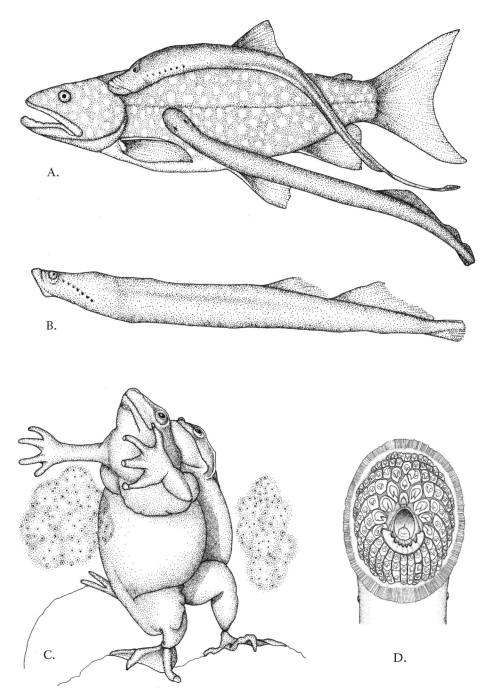

A.

B.

C.

D.

coordination. As it is, the surface layer turns to a milky soup as eggs and sperm cloud the water, maximizing the likelihood of sperm penetration. An incomprehensibly tiny amount of pheromone has been released with the eggs. These molecules, as if by magic, turn a school of fish into a procreative unit.

A male civet cat rubs his anal region on a tree. A stag demarks his territory with urine. The pheromone from the stag's anal gland is present in his urine. The pheromone retains its potency for long periods. This trait has survival value. If the male was to be unable produce a substance that retains the signal to females that they are invited to enter his territory (and tells competing males that they are unwelcome), he would exhaust himself endlessly squirting urine on the boundaries of his territory. He would become dehydrated from all this effort and expenditure of urine. No, the male must produce a substance that retains the attractant capability of the telltale urine. This substance is called *musk*. The Chinese government protects its herds of extremely valuable musk deer, *Moschus moschiferus*, the males of which are proficient producers of musk. The musk is sold to the perfume industry to enhance the longevity of its romance-inducing products. The more expensive the perfume, the more musk it contains—more magic molecules are hidden under the flowery odor.

When push comes to shove, we must face the fact that the nubile maiden is smearing herself with the secretion from the anal gland of a musk deer. The names given to scents reflect sexual psychology. Instead of a cologne labeled *Evening of Love*, the male attempts to attract a female with an attractant called something like *Stud*. One company even has the audacity to call its cologne *Musk*. None, so far, has called their product, *Eau d'Anal Gland*.

Humans have learned to utilize other animal pheromones. A male cockroach, universally hated except by female cockroaches, picks up the sweet smell of sex. He rushes to the source of the pheromone. It is a dark hole, the inner recesses of which hold the promise of procreation. This is a Roach Motel and he rushes to his doom, for it is laced with insecticide.

Pheromones are universal in the aquatic world. There are as many variations in chemically induced sexual behavior as there are mating rituals. A fecund female frog produces such a strong sexual scent that males from near and far are attracted to her. The successful suitor mounts the female and grasps her in a powerful embrace called amplexus. But wait, he has no penis! Having drawn the male to her, an instinctive copulatory sequence is initiated by the female. In a convulsive movement she releases a black and yellow mass of eggs. Detecting pheromones released with the eggs, the male deposits sperm on them in a complex chemical crescendo. His embrace has placed sexual openings in close proximity. Internal fertilization with its promise of maternal care has not yet evolved.

Usually pheromones are produced by the female to initiate the reproductive ritual. But one contrarian fish breaks the mold. The sea lamprey, *Petromyzon marinus* (found in freshwater, not the sea, in the United States), is a foot-long parasitic fish that came from Europe in the ballast water of a ship probably carrying cargo to a port in the St. Lawrence River. The lampreys escaped from their ship-borne prison and entered the Great Lakes, where they roamed freely, without natural enemies. Lampreys attack lake trout, attaching to the flank of the victim with a fiercely-toothed, funnel-like mouth, grinding away flesh until the host slowly dies.

The lamprey breeds in tributaries entering the lakes. In this case, it is the male who builds the nest and attracts the female. The pheromone has been described as Love Potion Number 3KPZS (3-keto petromyzonol sulfate).* A few molecules of this potent substance will attract females from far away. If a "Lamprey Motel" can be devised, the females could be lured to their deaths, and this problematic parasite would cease to destroy the trout fishery in the Great Lakes.

* Sharon Moen, "Romancing the Sea Lamprey," *Seiche*, March, 2003, 4–5.

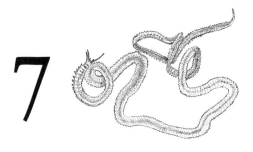

7

Skinny South Sea Sausages

BROWN-SKINNED MEN AND WOMEN slide their canoes into the roaring surf on this bright moonlit night. Beyond the surf tiny flickers of light, the moon's bright reflections, dapple an otherwise velvety sea. A vague white line off in the darkness demarks the insignificant surf weakly thrusting itself onto the protective fringe of reef offshore. The water's dark surface seems inviting instead of dangerous this evening. Beyond the reef a treasure lies beckoning in the distant darkness.

The full moon this October night signals the initiation of an ancient oceanic rite. Outrigger canoes slide through the warm water. Paddlers dip their broad blades into the sea with strong strokes. The canoe glides along until the man in the bow holds up his hand. Torches are lit, their flames revealing thin spaghetti-like animals swimming in aimless patterns. The Polynesians reach over with nets and bare hands, scooping up these fifteen-inch-long "sausages" and drop them into their mouths. Hardly a bite is necessary to pop the skin, and rich black eggs overflow onto the chin. Each participant smiles contentedly. This is a much-anticipated annual festival night. The hordes of juicy-looking swimming sausages are hardly reduced in number by this orgy of eating. At dawn, sated, the people return home with sacks full of eggs, to be replaced by a sky full of rapacious birds and silvery hordes of hungry fish.

How do fish and man know that this rich undersea ritual had commenced? Each year the Samoan palolo worm, *Eunice viridis*, a distant relative of the earthworm, reaches the culmination of its reproductive cycle and fills the sea with elongate egg-laden reproductive bodies. But close examination reveals no head on these fecund females. That accounts for their aimless paths as they crisscross each other in a living spaghetti soup of randomly moving wanderers.

All night long, at frequent intervals one or another swimmer pops open, spilling blackish beads into the water—eggs. Some of the aimless wanderers are males, and they release their sperm on the precise night that the eggs turn the near-surface waters into soup. The water turns gray as the white sperm mix with the black eggs. Fertilization occurs.

Each year on the same nights on the lunar calendar, male and female Palolo worms reach the culmination of the sexual cycle: fertilization and dispersion of the eggs. No matter how risky this ritual, a new generation is assured. The nutritious eggs supply energy to man and beast alike.* Sea birds and predatory fishes feast on the remaining "sausages," those that did not burst open in the night. But the danger does not end at the moment of freedom when the eggs pour out of their restrictive envelopes. A ferocious horde of zooplankton is waiting to pick off the freely floating bubbles of nutrition.

When spring approaches, the body of each female worm begins to change. Her posterior elongates and becomes fat with eggs—a rotund Siamese twin. The twin does not resemble its sibling. The front section of the female is thread-like. Its posterior twin is fatter, with wide, strong parapodia—paired paddles able to provide rapid movement. On the first days of the October and November fourth-quarter moons, the worms, having spent the year eating ravenously, begin to increase their activity. Suddenly a female rises from the bottom and plunges headfirst into the sand. Thus embedded, she twists her egg-laden posterior. It breaks free and

---

* Samoan recipes for experimental gourmets: fry with butter, onions, and eggs; spread raw eggs on toast, like caviar.

swims off. This process is called epitoky. Many worms perform epitoky, breaking off a posterior reproductive unit, the epitoke.

What initiates the process? Certainly tides and the currents they produce are part of the stimulus. Certainly water temperature contributes. The light of the moon filters down to the bottom in the clear-as-air water. It, too, must have something to do with initiating this annual cycle. But why this day? Why these hours? Why this moment? The answer is the same as that which accounts for the brief time span during which a school of mackerel spawns; when shrimp cast their gametes into the sea; when sea urchins and corals have their orgy of reproduction. The answer is pheromones (see chapter 6).

These substances, virtually ubiquitous in the animal world, are nature's signals. Only a few parts per billion dissolved in the sea can initiate some sort of animal behavior. They can induce agonistic (aggressive) behavior, in direct contrast to reproductive behavior, and "leave me alone" messages such as those released by female crabs to avoid prospective suitors.

᷍᷍᷍᷍ What to Do on Your Next Vacation in the Caribbean

Another, more accessible, cyclical reproductive ritual is exhibited by a member of the same worm family as the palolo worm. This worm, *Odontosyllis*, is a tiny thread-like animal hardly an inch in length. At our former lab in the British Virgin Islands there was a long, T-shaped pier. One evening my graduate student suggested that we go to the end of the pier to look for mating *Odontosyllis*. "It's one thing to read about obscure phenomena, and another to actually try to find them," I thought, shaking my head skeptically in reluctant agreement. We reached the end of the pier. I waited, enjoying the lovely tropical breeze and the star-studded sky. Then he shouted, "There she is!" Incredulous, I ran over to see for myself. A bright green, luminescent halo floated slowly toward the pier. Inside the halo was a startlingly bright, tiny crescent—the female worm. The female of this species is capable of releasing a cloud of bioluminescent mucus to attract males. Whether or not pheromones are involved in the attraction is unclear, but it is likely that they initiate the reproductive process.

If you want to see this fantastic phenomenon, search for *Odontosyllis* in the Caribbean on the first days of the third-quarter moons in June and July.*

ᔐᔐᔐ How Do Fireworms Mate? Slowly

The fireworm, *Hermodice carunculata*, is a beauty. It can be a foot long, and its fat olive body is bordered with a red and white fringe. It is one of those colorful, obvious animals that flaunt their beauty in the turtle-grass beds of the Caribbean. An eager student, on the first day of class (before I had a chance to issue warnings), brings me his first specimen, a fireworm, expecting praise. Instead, he gets a look of horror. Unlike fishes, which avoid bright, conspicuous prey, students venture in where fishes fear to tread. As I look with consternation at the worm writhing on the student's palm, I explain that the glistening white borders of the worm are thousands of glass-like, minute barbed needles covered with poison. To touch the worm is to be impaled with dozens of these minute setae that remain embedded because of their barbs. The poison causes an inflammation, and the student is sidelined for a few days until the setae work their way out.

Snorkeling off Barbados, I came upon a foot-long fireworm. It was moving so purposefully that I decided to follow it. Don't ask me how I interpret a worm's movement as purposeful, but some sixth sense told me to follow the worm. To my surprise, I came upon its destination. There on the bottom was a soccer ball-size mass of the most gigantic fireworms I have ever seen. I was shocked at this writhing, living sphere. It boggles the mind to see this mass of mating worms. I think of an orgy of mating porcupines,

---

* A visiting professor from Chicago literally became ecstatic at seeing this phenomenon. I have never seen such ecstasy outside the bedroom. She brought classes to our lab in Jamaica for many years after that epiphany.

PLATE 7

A. Dawn after a spawning event. Birds and fishes are feeding on the remaining epitokes. Undulating black lines are swimming epitokes; straight lines are females that have split open to release their eggs. Lower are females surrounded by clouds of eggs.

B. Samoan palolo worm, *Eunice viridis*, in non-reproductive state. Up to 12 inches long, the thread-like animal spends its life in tubes in coral rubble on outer reef, emerging only to release its epitoke.

C. Female Caribbean bioluminescent threadworm, *Odontosyllis enopla*, surrounded by a halo of bioluminescent mucus that attracts males. Many males approach the luminous cloud, emitting spark-like flashes. Circling the glowing halo, they emit sperm simultaneously as the female emits eggs. The brightly glowing female is ¾ of an inch long. The threadworm is in the same family as the palolo worm.

D. Caribbean fireworm, *Hermodice carunculata*, can be olive, tan, or brown, bordered by thin red lines. To 10 inches. Edged with thick band of white spicules (not visible). When touched, the barbed, toxic spicules enter the skin. Fireworms gather into balls of many individuals to mate.

E. Epitoky in the marine syllid worm *Autolytus*. The parent worm is budding off epitokes. Unlike palolo worms, these epitokes are duplicates of the parent worm. Epitoky is a form of asexual reproduction, budding off clones of the parent. But epitokes contain eggs and sperm for sexual reproduction, providing for genetic diversity. The worms are 1 inch long and live in tubes on cnidarian polyp colonies upon which they feed. Male epitokes are white with sperm, females golden with eggs. The mating dance takes about twenty minutes as the male epitokes swim rapidly around passive female epitokes. The male extrudes mucus threads carrying sperm, wrapping them around the female. She releases hundreds of eggs that are fertilized immediately and stick to her body to form a golden halo. She broods the eggs for a few days, releases them, and dies.*

* After Vicki Pearse and Ralph Buchsbaum, *Living Invertebrates* (Palo Alto, CA: Blackwell Science, 1987), 412.

PLATE 7

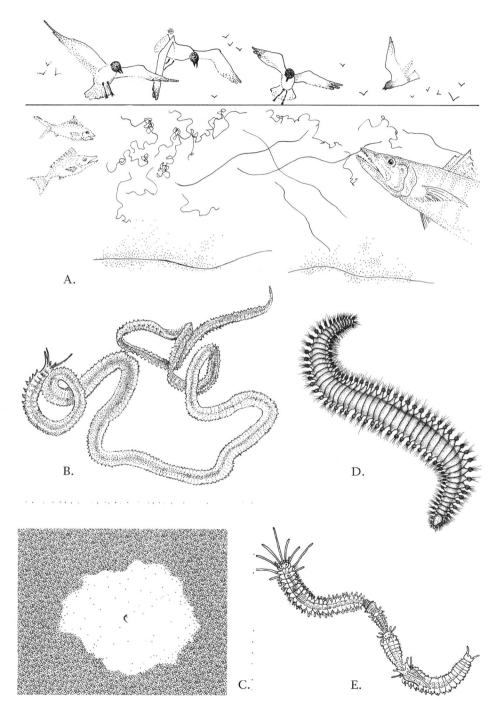

A.

B.

C.

D.

E.

and it creeps me out. How do these worms get together to copulate? Either they have some control over their setae, or else they end up like a pincushion after mating and the poison has no effect.

What stimulus caused the worm to move purposefully toward the mating ball? Of course, I detected nothing in the water. But like their internal analogues, the hormones, the external pheromones affect behavior at almost infinitely small concentrations. I wondered: Was something in the water causing ME to swim towards the mating worms?

8

The Only Male
Reproductive Organ
with a Name

THE MALE SQUID approaches the female and jealously guards her from his competition, all the while undulating his fins and darting to-and-fro. Flaming bands of red and lavender pulsate down his tubular body. As the mating dance becomes more frenzied, his color darkens to purple, then blanches; he becomes almost invisible. Accepting his charms, the blushing female turns pink, then lavender, waves of color rippling across her body. At the moment culminating the mating dance, he removes a sperm packet from his spermatophoric gland with his fourth right arm and places it into the mantle cavity of the female. She will save it for later use to fertilize each egg when it is produced.

The male will fight off any competitor. It is a matter of life and death. Somehow, in the depths of his non-thinking mind, there is an evolutionary cue to protect his genes, which are now safely stored in the female's mantle cavity. Should another suitor appear, he sallies forth to accept the challenge, changing his nuptial colors to become dark and threatening. The battle is joined in a flurry of twenty flailing arms and two menacing beaks. Should the competitor gain the advantage, the original suitor slinks off. The winner approaches the female. The first thing he does is to insert his siphon into the female's mantle cavity to blow out the sperm packet of his competitor. The genes that dictate his reproductive initiative (those same genes that cause each male and female to perform the mating ritual) are somehow programmed to eliminate the genes of his competitor. He is unaware that this process selects the strong and eliminates the weak.

❦❦❦ Sexist Parasite

On a research cruise in the nineteenth century, a scientist was routinely dissecting a live squid on the deck of the ship. He found a strange, wiggling "thing" in the female's mantle cavity. Examining other females, he saw the same "thing." He pondered over the fact that this puzzling writhing "organism" was found only in females. Excitedly, he pronounced it one of the few parasites that infect only females, and created a new genus for it: *Hectocotylus*. Many years later someone discovered that *Hectocotylus* was merely the spermatophore-bearing tip of the fourth right arm of the squid that had broken off in the mating frenzy. Somehow the name stuck and the tip of the arm used to transfer sperm has its own name. Although many couples have pet names for the male member (Mozart called his "my little soldier"), there is only one male reproductive organ with an official name: hectocotylus.

❦❦❦ Squid Sex

Copulation of the squid is a face-to-face encounter in which the lovers hug each other in a ten-armed embrace. Sperm are stored in a club-shaped spermatophore, at whose apex is a removable plastic-like cap. Eggs are produced in the ovary and move through a tube to the oviducal gland that covers them with albumen. On copulation, the hectocotylus twitches and gyrates inside the female's mantle cavity, as if it were an autonomous organism. In some species, it removes the cap on the spermatophore and glues the sperm to the female's gonopore (reproductive opening). In other species, the hectocotylus breaks off and uses its suckers to adhere to the female's body to release the sperm from the spermatophore as needed. The sperm enters the female's gonopore. Each egg is fertilized in turn. The female then covers the egg with a rubbery egg case and glues it to a rock or seaweed, where it is vulnerable to predation.

The mating process of the cuttlefish, *Sepia officinalis* (a closely related squid-like species), has been studied in detail.* First, the male "rushes" the female as if attacking. In a moment he pulls her to him and they are face to face. Mating takes an average ten minutes, six of which are spent violently blowing other males' spermatophores out of the female's mantle. Then the male uses his hectocotylus to remove a large bundle of spermatophores from his funnel and place them deep into the female's mantle cavity. For an average of three minutes, he grinds them in a mortar and pestle movement, freeing the spermatophores from the bundle. Many adhere around the female's paired reproductive openings to eventually release their sperm.

The male guards the female (and his sperm) for an average of ten minutes before both wander off to mate again. Any threat is met with a fearsome color change to the powerful zebra pattern, a stark, jagged black-and-white striped configuration.

The hectocotylus often is a spatula-like modification of the tip of the fourth arm that often looks little different from an ordinary arm. Sometimes it has a smooth, suckerless tip. In some squids the suckers are absent but their stalks remain as flexible fingers to manipulate the spermatophores. Some species have subtantial modifications for spermatophore transport.

In cuttlefishes, courtship is initiated by the female as in most animal relationships. She releases a pheromone—an alluring hint of impending sexual union—and all nearby males come courting. She is choosy. The males approach and she jets away from them. If one suitor becomes particularly insistent, she "inks" releasing a murky cloud of black ink as a repellent.

* T. Hanlon, S. A. Ament, and H. Gabr, "Behavioral Aspects of Sperm Competition in the Cuttlefish, *Sepia officinalis*," *Marine Biology* 134 (1999): 719–28.

PLATE 8

A. LONG-FINNED OR COMMON SQUID, *Loligo pealei*, has captured a shrimp with its pair of long tentacles. Its other appendages are called arms. Whitish with lavender sheen when not provoked, up to 16 inches long.

B. TWO MODIFICATIONS OF OCTOPUS ARMS FOR SPERM TRANSFER. One octopus species has a paddle-like hectocotylus; the other has a filament-like one. In many species of squids and octopods, the hectocotylus is indistinct.

C. MALE OCTOPUS coyly inserting his hectocotylus into the mantle cavity of the female. This occurs after an elaborate ritual involving extremely rapid color changes. Octopods and squids use sacs of three primary colors of pigment called chromatophores. These can be expanded and contracted under the control of the nervous system, permitting changes as rapid as nerve impulses. As an octopus crawls over the bottom, its color changes so rapidly that it is constantly camouflaged.

D. TWO MALE SQUID-LIKE *Nautilus* in combat. Waves of color flow down the body, eventually producing a fearsome, aggressive zebra pattern. The dominant *Nautilus* (upper) exhibits a strong pattern. Color is fading in the submissive loser. If the challenger wins, he will squirt a jet of water into the mantle cavity of the female, washing out the sperm of his predecessor.

E. MATING SQUIDS. Excited animals produce an ever-changing bouquet of color. The male uses his hectocotylus to insert a spermatophore (sperm packet) into the female's mantle cavity. Some spermatophores have caps on them and are uncapped in the mantle cavity. Some writhe around like living organisms, releasing sperm with every egg shed by the female.*

---

* After W. D. Russell-Hunter, *A Life of Invertebrates* (New York: Macmillan, 1979), 43.

PLATE 8

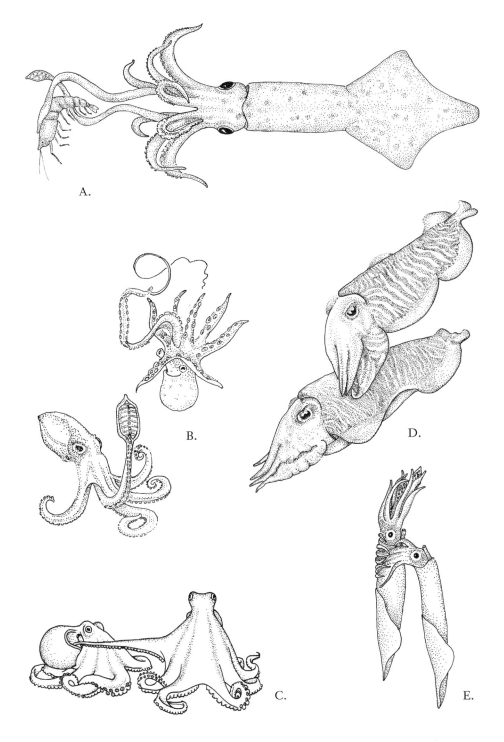

A.

B.

C.

D.

E.

Finally she succumbs. In fact, she succumbs a lot, mating repeatedly, leading to the process whereby males blow out their predecessor's sperm.

~~~~ The Top Predator

The mating process of the squid is remarkably effective. Huge, uncountable schools of long-finned squid, *Loligo pealei*, have been recorded. In some instances in the early twentieth century, it took hours for a school to pass under an anchored boat. Massive numbers of squid, in an annual mating ritual, enter California bays to deposit flask-shaped egg cases, each containing one embryonic squid. Schools of sharks and other piscine predators descend on these submerged fertile fields in a feeding frenzy. Despite the depredations, millions of juvenile squid hatch from the protection of their two-inch-long translucent havens. They are too large to be preyed on by the denizens of the planktonic jungle, but some of the predatory fishes linger, taking their toll. Natural enemies reduce their numbers, but the surviving millions are more than enough to repeat the mating ritual the next year.

But enter the most voracious predator of them all—man. The knowledge that squid is good to eat has long been apparent to the Japanese, Koreans, and Mediterranean peoples. They have traditionally caught squid by jigging small hook-laden lures in mid-water. Modern Korean squid fishermen have improved on the traditional jigging technique, using mechanical winches that jig the lures up and down. At a dialed-in weight, the jigging machine pulls up the line, whipping the impaled squid around a wheel, throwing them off the barbless hooks onto the deck where they are swept into the fish hold and iced. Very efficient.

But good old American efficiency has responded to fashion. A new-old appetizer has entered haute cuisine, calamari—Italian for fried squid rings. The Greeks and Portuguese call it "calamaris" in their restaurants, but the result is the same: a huge demand has caused a response from the American fishing industry. Artisanal hand fisheries have given way to huge factory ships that locate the gigantic schools of squid with sonar and surround them with the gaping maws of half-mile-wide nets towed behind the ships. In one moment, the huge school has disappeared, only to reappear as succulent fried rings made of the tubular muscular mantles of the squid.

Just as red drum (redfish), *Sciaenops ocellata*, became commercially extinct when a New Orleans chef published his recipe for blackened redfish, squid of the Atlantic coast have reached dangerously low populations. The frozen squid blocks available in your supermarket for a few dollars are Pacific species. The squid, whose life cycle encompasses just one year, has become as threatened as its taste has become fashionable.

9

Living Lance

A HUNDRED MEN pull a thick, rough rope, occasionally stumbling on the slippery sand, backs bent as powerful testimony to their labors. The women of the village mill around at the water's edge. Occasionally one or two jump up out of the throng to better see the sweaty scene. All loudly shout encouragement in that strange mixture of Dutch and an African dialect called Papiamento. A puller courteously steps back, offering me a three-foot section of inch-thick rope on which to pull. It fills my hand with its rough, splintery width. The rope extends out to sea in an almost invisible crescent, partly delineated by a rowboat, tiny in the distance. Its occupants, two tall, skinny men, teeter excitedly, exhorting the pullers to keep the net tight. Its indistinct length stretches a quarter-mile parallel to the shore, making that difficult.

The riot of color created by wet, shiny black skins contrasting with orange and red bathing suits is a living Haitian primitive painting. The intensity of this powerfully sensory scene is enhanced by the white-hot sunlight and dun-colored sand, framed by dark green palm trees. Each pull on the mile-long rope attached to the net is accompanied by a chorus of grunts as, backs doubled over and glistening, taut wet muscles (and my beet-red skinny arms) yank in a few feet of the heavy net at a time.

At last the silvery catch becomes visible, whipping the aquamarine water to a froth. The villagers, unable any longer to contain themselves, leap into the slippery, silvery mass, plastic bags tucked into the waistbands of their bathing suits.

❦❦❦❦

The catch was the sharp-toothed needlefish, *Strongylura notata*, as unlikely a delicacy as one might find. Its foot-long body, thick as a banana, seems to be all toothy jaws. It looks like a miniature barracuda, only more ferocious. Despite its green-hued flesh, this is the poor man's delicacy—the most easily caught edible fish available to the inhabitants of fishing villages in the tropics, because it swims close to the surface in schools of a thousand or more.

On another, rougher occasion, Bob, our enterprising boat captain and a commercial fisherman-cum-cattle farmer, happened on a school of needlefish in the shallows near St. Ann's Bay, the closest town to our lab in Jamaica. He and his son and brother surrounded the school with two boats using a floating net off a beach about a mile from town. Someone noticed the characteristic pattern of two boats with a crescent net between them, and the news was out. About fifty locals lined the beach, shouting encouragement. By the time Bob could get his boat in position to tighten the net around the school, all fifty people were in the water, shouting, laughing, and hauling in fish—fish in pockets, fish in skirts, fish in pants. If someone was gathering fish too aggressively, Bob or his son would pick up an oar and make a dash for the villain, only to see someone else stuffing a big shopping bag with the squirming delicacies. In the end, Bob filled his boat with a fraction of his silvery harvest and had to be satisfied.

We use needlefish for our own educational purposes. Needlefish-catching contests have long served as ego boosters on nighttime expeditions with our class in tropical marine biology. As the great equalizers, they provide a silver reward for rich man or poor, and in our case, an ego boost for nerd or macho man.

On our way out to the octopus-hunting grounds at night, we wade thigh-deep in the inky shallows, waterproof flashlights creating small circles of light that attract silversides and a variety of planktonic animals. Attracted in turn by the silversides, needlefish suddenly appear as ghostly green flickers fading in and out of the penumbra of light. You stop, hoping one of the other people stretched out in a glowing irregular parade off into the distance won't blunder into your area, scaring it away. More often than not, it disappears with a flick of its tail, and you are out of luck. But if you are lucky, it comes within range and you grab at it.

Amazingly, you are standing there in a circle of flashlight beams with your hand raised high, a needlefish flopping in your fist. You utter a primeval shout of triumph. The balmy night and the soft tropical breezes have reduced you to your ancestral hunter-gatherer persona. A few days ago you were in Chicago or on Wall Street. Now you are standing thigh deep in the black sea waving a live fish and howling in triumph, surrounded by your tribe of cheering yuppies.

One year, the contest climaxed with a competition between the two personality extremes in the class—the strong, garrulous type and the weak, silent type. Both were tied with four fish. The more aggressive guy knew that a bottle of Gold Label rum and an honorific certificate at the going-away party would be the reward, so he shadowed the underdog (whom we were all rooting for). Suddenly, at the same time both saw a hapless needle-fish paralyzed by the glare of their flashlights. The nerd was closest, but the aggressive guy butted him out of the way. Both lunged—there was a huge splash—and the nerd triumphantly surfaced with the fish clutched in his fist! His face glowed as I presented him with his certificate as Olympic Champion Needlefish Catcher and his bottle of Gold Label rum at the going-away party. This record (five fish) has never been broken. (Clutching the needlefish does not harm them. When placed in the water immediately after being waved in the air, they disappear instantly with a flick of the tail. I have never seen a needlefish harmed by this human contact, but I have seen a human ego saved.)

Seeking Safety in Seething Seas

The upper sunlit layer of the ocean is a thick soup of plankton, jungle-like, where predator and prey dart about in their dance of death. The larval needlefish must face this awful threat.

The water is green with minuscule plant-like phytoplankton. They use chloroplasts to manufacture sugars and the precursors of proteins. Some keep afloat using glass-like siliceous spicules, some swim using sperm-like flagella. Some calcareous, exquisitely pitted, microscopic spheres maintain position in the water column using buoyant, energy-laden oil droplets. It is this energy capsule that sustains their predators, the often invisible zooplankton. The phytoplankton are lowest on the trophic pyramid. Those that eat them, the herbivores, take an astonishing variety of shapes, from tiny larval clams that vacuum phytoplankton into their microscopic mouths, to minute, worm-like animals that use ciliated fans to suck in the energy-rich phytoplankton, to ciliated swimmers, to the ever-present copepods. At the next level of zooplankton are vicious-looking micropredators exemplified by *Sagitta*, the arrow worm, a minute monster with fierce fangs used to snare passing invertebrate larvae, fish eggs, and small copepods. Ciliated and flagellated zooplankters whirl and stagger by, to be engulfed by even larger planktonic carnivores, among them the larval needlfishes.

The minute larval needlefish, poorly endowed with yolk, must capture one of these zooplankters within hours of birth in order to survive. Eggs have been cast into the sea by an adult female needlefish compelled by evolutionary happenstance to produce thousands of eggs, dividing all her spawning energy among their huge numbers. Consequently, the tiny larval needlefish lacks the strength to search for prey. A bumbling hairy football—a jellyfish planula larva—or that ubiquitous microcrustacean, the copepod, must swim or float nearly into its mouth. So immature is its nervous system that it literally must wait for its prey to brush against it accidentally. In these lush feeding grounds the surviving newly hatched needlefish larva grows to foot-long maturity and joins with others of its year-class in a school.

Safety in Numbers

Needlefish form dense schools, darkening the water with their teeming masses. At any moment the black swarm is punctuated by a burst of light as a random fish flashes its silvery sides like a firefly's brief spark in the dark sky. The school, a huge dark ameboid shadow, presents a formidable

PLATE 9

A. A NEEDLEFISH, *Strongylura notata*, feeding on a dense school of silversides near the surface. Behind it is a ballyhoo, *Hemiramphus brasiliensis*, similar in size and shape to the needlefish. Needlefish are silvery, greenish on top, up to 25 inches. How can a school of needlefish a quarter-mile wide find enough prey if the huge school remains intact? A catch 22: stay together in the safety of the school or disperse to find food. Dispersal into schools of six or eight occurs at night.

The ballyhoo is included as a curiosity. Its family name, Hemiramphidae, means half-beak. Needlefishes and ballyhoo do not ordinarily school together. Despite their bizarre appearance, ballyhoo are predators. About half their food is small fishes and the rest sea grasses.

B. JUVENILE NEEDLEFISH chasing a copepod. These one-eyed microcrustaceans provide important sustenance for juvenile predators in the sea because they are the most common marine microscopic animals. It is unlikely that inch-long juvenile needlefishes would be able to survive without the ubiquitous copepods.

C. SCHOOL OF SILVERSIDES crowding together, creating a tunnel. Instead of engorging on the abundance of prey, predators often burst out of surface, confused and hungry, unable to focus on an individual silverside. Divers often find themselves in the same sort of tunnel, surrounded by a school of fleeing silversides.

D. HAND OF A STUDENT clutching a squirming, slippery needlefish. The fish are attracted to zooplankton that come to the halo of light produced by a flashlight. Zooplankton dart through the halo from every direction. The needlefish seems to be confused by the glare and profusion of food, and it can be grabbed. Most often it simply disappears with a flick of its tail.

E. HOUNDFISH, *Tylosaurus crocodilus*, a large relative of needlefish. Dark blue above, silvery white flanks, dark blue stripe along sides. To four feet. Due to its propensity to leap out of the water when stressed, this fishy missile is truly a living lance.

PLATE 9

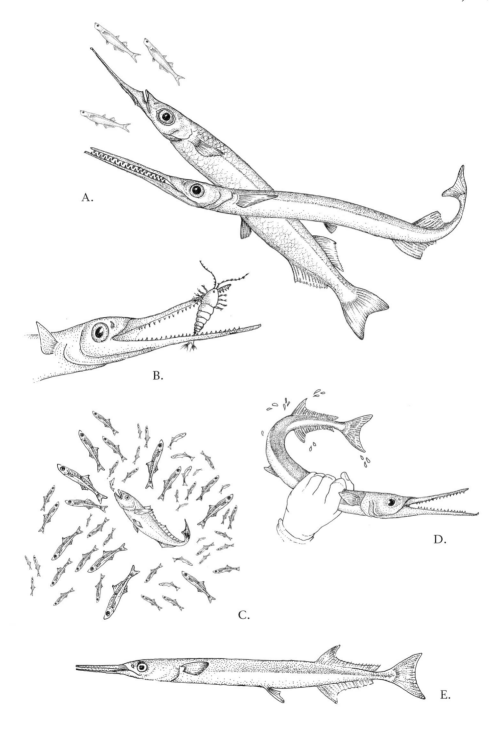

A.

B.

C.

D.

E.

threatening presence when viewed from underneath. The random flashes of reflected light further confuse the picture. The dense shadow in the water undulates in a massive flow going nowhere. Disturbed briefly by a hungry predator's dash through the school, the edge of the school fragments, the fringes a scattering of terror-stricken needlefish. The predator charging voraciously into the middle of the school finds itself in a tunnel of frantic needlefish pushing against one another in a convulsive contraction. Often, at the end of the tunnel the predator ends up empty-mouthed, sometimes bursting through the water's surface, tossing end over end to fall back with a splash.

When a school of needlefish is attacked by hungry tuna, the surface is whipped into a froth of leaping needlefish and their predators. Above this melee swoop frigate birds, gulls, and pelicans waiting to take advantage of the scattered needlefish bursting from the school. Underwater, hungry predators at the edge of the school attack the now-vulnerable needlefish scattered by the attack. Once separated from the amorphous mass, the needlefish are easy prey, defenseless and readily picked off. The school takes the place of terrestrial sanctuaries—forests, burrows, and rocky havens. In the absence of these, in the featureless open ocean, the needlefish has learned that sanctuary lies in closely packed schools; individuals are at risk. (On land, wildebeest evade the lion's charge by herding together, flank touching flank. Confused by the huge mass rumbling by, the lion cannot zero in on one prey animal. No wonder lions are frequently unsuccessful in their hunts).

Needlefish live close to the surface of the sea. Their backs are greenish, dissolving into the color of the water, virtually invisible from above. Silver flanks complement the watery greenish upper surface of the fish, mimicking the silvery reflection of the water's surface when viewed from below. But sometimes the protection of the school, dark with densely packed fish, fails, as it does when a swordfish or marlin attacks, blindly whipping its "sword" through the school. Its stunned prey, crippled, sink slowly as the frightened school escapes. The swordfish returns soon after to engulf the injured.

When an individual needlefish finds itself isolated from the school, it resorts to its last line of defense. Needlefish are great jumpers. They can skim over the surface for many feet, evading a predator too heavy to follow its spectacular escape through the air. Closely related species, the houndfishes *Tylosaurus crocodilus* and *T. acus*, can reach a length of four feet. They are called living javelins, as they burst from the water like an explosion of silver spears when frightened. One fisherman tells the story of being impaled by a hysterical houndfish, its beak-like jaws penetrating through his calf muscle.

The ultimate defense mechanism of the needlefish and their brethren—not the individual but the species—is their fecundity. A 52-inch, 10½-pound houndfish female produces an average of 25,000 eggs, each as large as this "o". These eggs are scattered by the females in the open sea and must depend on the massive release of sperm produced by the males at the same time. The common needlefish is an inshore species. Its eggs are embellished with a thread-like filament to allow for attachment to floating debris. The eggs hatch, releasing tiny fry that grow into blackish, twig-like, inch-long juveniles. As they grow larger, they mimic their hideaways, floating flotsam or sea grass, emerging to gorge themselves on zooplankton. Eventually, growth forces them to forsake the flotsam and they coalesce into a shadowy, undulating school ruffling the water's surface. Hungry predators (including man) await.

But the odds of survival of the species are good, for only two fish of the thousands of eggs cast into the water by one pair of fish must survive to reach adulthood to replace their parents in the ongoing stream of life.

10

Role Reversal

THE SEAWEED sways to-and-fro with the ocean swells, yellow-tan fronds glistening in the sunlight. The movement is so regular as to be hypnotic. Bulb-like swellings blister the flat, strap-like *Sargassum* strands, accentuating the flowing movements. Seen from above, their spikey projections poke out of the ocean's flat surface, making the ten-yard-wide mass appear as a golden carpet. The fluidity of movement is interrupted. Barely discernable, the outlines of two shadowy, miniature horse-like shapes can be picked out of the seaweed jungle. Their color exactly matches that of the seaweed, and the sinuous movements of their dorsal fin "manes" replicate the flow of the sea.

From their perches on fronds of seaweed, wearing their nuptial attire, the newly colorful couple unwind their prehensile tails and begin the reproductive ritual. The male seahorse "bows" to the female's "curtsy." The pair do-se-dos side by side, tails entwined in a slow, stately aquatic square dance. Then they grasp a vertical seaweed frond and begin to twirl around it. Faster and faster goes this maypole dance. If the pair is compatible, this nuptial maneuver prepares the gonads for the culminating act. At the climax of the dance the frenzied stallion approaches the waiting female—but here the process becomes unique. Instead of an active role, the male remains passive, as, bodies entwined, they momentarily bump bellies. During that fleeting moment the female deposits as many as several hundred eggs into *a pouch in the male's abdomen*.

The courtship ritual is concluded. In the male's pouch, sperm are released to fertilize the eggs. Here they develop for about ten days until the male's pregnancy reaches term, his pouch swollen almost to bursting. The pressure of the swelling ejects a cloud of tiny, miniature sea horses. *The male gives birth!* He has brooded the juveniles until they are capable of functioning as free-living organisms, vaguely similar to the brooding of mammals.

❧❧❧❧

The difference between the brooding process of terrestrial horses (viviparity) and that of seahorses is that the female horse uses a placenta as a means of providing for the needs of the fetus. Seahorses provide protection only, not nourishment (ovoviviparity).* Birds lay their eggs in nests, and the egg nourishes the fetus with a rich yolk. The egg is brooded outside the body. Caring for the fetus outside the body is called oviparity.

By contrast, most fish species provide no protection for their offspring, casting thousands of eggs into the threatening sea. In comparison to the brooding process that protects birds' eggs, fishes use a primitive version of oviparity. Those eggs that survive the ravenous predators of the planktonic jungle hatch into helpless larvae that can hardly move. They subsist on yolk for a period of hours and then drift along ill-prepared to defend themselves, much less eat. Their ranks are decimated. Not so the seahorse. To be brooded gives the quarter-inch juvenile a tremendous advantage. It is released from its paternal haven too large to be vulnerable to most microscopic planktonic monsters. In fact, like a zebra foal, moments after birth it is able to escape predators and find food.

Seahorses, miniature aquatic replicas of terrestrial horses, have evolved a long horse-like snout and curved abdomen resembling the noble neck of

* There is a dispute concerning an additional role. It has been suggested that the skin lining the pouch helps maintain the pH (acidity) of the larval environment or produces a milk-like nutritive secretion, or becomes a spongy, placenta-like nutritive membrane.

an Arabian steed. At the end of the snout is a small diagonal mouth that sucks in unsuspecting microcrustaceans. The mouth closes on the prey with an audible snap. Seahorses do not have scales. These have been replaced by hard bony scutes, defense against small predators. Their primary protection is camouflage. They cannot escape large predators because their only means of propulsion is an undulating, filmy dorsal fin with which they slowly paddle themselves around.

✤✤✤ Seahorse and Sex

Twenty species of seahorses are currently listed in the World Conservation Union (IUCN) Redbook as endangered or vulnerable. The usual reason, these days, that animals become extinct is anthropogenic (caused by human activities). Human-related causes for species extinction are often subtle, complex, and open to controversy. Examples of these indirect influences on population survival are habitat destruction and global warming.

However controversial the issue, the near extinction of seahorses is clearly attributable to man, not in the usual vague sense but uniquely human-to-seahorse. There is an ancient tradition among old Japanese men that impotence can be cured by a potion made of dried seahorse (Japanese folk Viagra). The formula is: take ten small or five large fresh or dried seahorses, grind them into pulp or powder, mix with a cup of saki, and drink an hour before attempting sexual activity.

Imagine the wealth that could be attained if someone were to take an attache case filled with iced seahorses and visit every whorehouse in Tokyo to sell these wares to each elderly gentleman come upon.

✤✤✤ Feeding the Babies

Unfortunately, seahorses are considered cute. They are among the most-desired fishes enjoyed by aquarium hobbyists, which number in the millions. In the only census of seahorse harvesting, the state of Florida recorded a tenfold increase in capture from 1990 to 1994, to 112,000 individuals. It has been estimated that the world's voracious harvesting of these animals exceeds 20 million seahorses.

To supply enough seahorses to satisfy the needs of Japanese seniors and counter other serious threats to seahorse survival, we have developed a technique for farming seahorses. They are difficult to culture, despite their birth as miniature adults, because they are notoriously picky when it comes to taking their first bite of food. Normally when we raise larval fishes, we feed them the smallest multicelled organisms, rotifers (microscopic vacuum cleaners with two whirling "brushes" of cilia that sweep phytoplankton into the mouth). These we culture until they fill a container in uncountable numbers. Too small to be seen by humans, their soupy masses are offered to juvenile seahorses in volumes that make it certain that many rotifers are available to each juvenile. To be sure that we are providing adequate nutrition, we dip the rotifers into a solution of vitamins and fatty acids, thus enriching them. Still, the infant seahorses will not thrive on the rotifers and, gradually, the culture tanks become littered with tiny carcasses.

Another technique, which should work with these quarter-inch juveniles, is to feed them the earliest larval stages of the brine shrimp *Artemia*. Brine shrimp are bite-sized morsels favored by most small aquatic animals. Over millions of years of evolution they have developed the ability to produce prodigious amounts of egg-like cysts. These are washed onto the shores of hypersaline bodies of water like the Great Salt Lake in Utah in such numbers that entrepreneurs bulldoze the sand into trucks, strain the cysts from the sand, and package them in coffee cans to be sold. Each can of cysts may sell for as much as $75. The high price connotes the desperate need of the aquaculture industry to feed them to larval fishes and shrimps.

The energy-rich, yolk-filled early brine shrimp larvae are produced by placing the cysts in a cylinder of bubbling brine for twenty-four hours until hatched. Then the first stage larvae (called a nauplius) are fed to the juvenile seahorses. The nauplius has no mouth and cannot eat. To survive through metamorphosis to the feeding stage, it depends on highly nutritive yolk. This makes brine shrimp larvae excellent food. But alas, only a small number of seahorse juveniles feed and survive on this traditional food. Successful culture using conventional brine shrimp larvae is rare.

Why? What mysterious stimulus will cause the juvenile seahorses to feed? Solving this question occupied us for months. The answer, we found, is that the temperamental juveniles won't eat unless stimulated by a distinctive erratic swimming pattern of the prey. Finally, after many frustrating experiments, we found that the only prey organism with the necessary stops, starts, and angles to its swimming pattern is the copepod, the dominant herbivore of the planktonic world.* Its dense populations dwarf its terrestrial counterparts, rabbits and mice. It is eaten by juvenile fishes and planktonic carnivores, snatched from the water by herrings by day and soldierfishes by night. Gigantic basking sharks and whales strain them from the sustaining sea. Few aquaculturists have been able to culture copepods, so they must be collected from the ocean.

Finally the breakthrough! We fed wild-caught copepods to the juveniles. They thrived. Gradually we weaned them onto newly hatched brine shrimp. After a few weeks, the juveniles are large enough to eat another wild planktonic animal, the shrimp-like mysid, which can be bought in frozen blocks. Eventually the young seahorses grow to about an inch long and they can then be fed adult food, finely ground shrimp.

Our plan to substitute cultured seahorses for wild ones can satisfy the demands of the aquarium trade and the Japanese impotence-related demand in the short term. A long-term program to insure the survival of a species is called stock enhancement. It is our role as scientists to share the information on seahorse culture we have generated by publishing the procedure in journals that are read by colleagues all over the world. They will modify the technique to their own environmental conditions and use it to produce local species in large numbers, releasing them into the wild to augment local populations.

For example, the coral reefs off the Philippines have been virtually denuded of their indigenous species of seahorses, and three species common to the reefs off Australia are close to regional extinction. One of these endangered species is the spectacular leafy sea dragon, *Phycodurus eques*. This gorgeous fish is ornamented with long, leaf-like papillae extending from the body, matching the frilly seaweeds among which it lives. In nature, among the seaweeds, the eight-inch animal is virtually invisible. But in an

---

* This research was performed in my laboratory by graduate student Todd Gardner.

aquarium, it dominates with its exquisite beauty. Frond-like papillae flowing, it moves with stately grace. Its beauty is its undoing. It is much sought after by aquarium hobbyists, a good specimen selling for as much as a thousand dollars. Naturally, such living gold is in high demand. The sea dragon's slow movements prevent escape from experienced collectors, who have an uncanny ability to pick it out of a seaweed patch, near-perfect camouflage notwithstanding.

✤✤✤ Collateral Damage

Can seahorse populations withstand the sexual appetites of elderly Japanese gentlemen and the aesthetic demands of aquarists? To complicate things, disastrous "collateral damage" accompanies the quest for seahorses and other tropical fishes. In the Philippines, inhabitants of whole villages devote themselves to gathering fishes for the aquarium trade. They often utilize high tech collecting techniques. Early in the morning the men of the village paddle out to the coral reefs. Putting on their scuba gear or snorkels and carrying hand nets, they jump into the water. On their belts are small plastic squirt bottles filled with cyanide solution. Upon seeing a choice specimen, they close in. The frightened fish darts into a crevice in the reef. The fisherman squirts the cyanide solution into the hole. In a few moments the fish floats out belly up. When placed in the fisherman's resuscitation tank, it rights itself and erratically swims off. Those that survive are shipped off to New York, Berlin, or London. But no matter how healthy they appear, they often die in a day or a week from the residues of the poison.

The coral polyps subjected to this moment of mayhem are more sensitive than the fishes, and soon a halo of dead white tissue surrounds the crevice where the fish hid. White blotches sully the reef, evidence of the incidental death of the coral. Near the village the blotches coalesce and the reef dies. Radiating out from the dead reef are other reefs, ever farther from the village, where the collectors are forced to forage for fish. These reefs exhibit the telltale blotches signifying their doom.

PLATE 10

A. LEAFY SEA DRAGON, *Phycodurus eques.* Yellow, tan, or green, to 8 inches. The sea dragon's head is pointing to the right. Look for the eye and long snout. Color matches the golden sargassum weed, *Sargassum fluitans,* its primary habitat (see the two fronds of sargassum weed at the surface). This species of golden algae floats near the surface in huge rafts, never attaching to the bottom, and provides a fine refuge for the leafy sea dragon.

B. LINED SEAHORSE, *Hippocampus erectus.* Brown to red, to 6 inches. Dark lines run diagonally across the head, dark vertical streaks along the body. Bony scutes and camouflage protect it from predation. Swims by undulating its dorsal fin, too slow to flee predators. Often clings to seaweed with its prehensile tail. Common on the Atlantic coast.

C. COURTSHIP AND MATING DANCE OF THE SEAHORSE. After dawn of the first day, male and female signify readiness for mating by becoming brightly colored. The couple clings to a blade of seagrass and circles like merry-go-round horses. At frequent intervals they swim in parallel to a blade of grass. The male then jackknifes vigorously, pumping water in and out of his pouch. On third morning, the female's trunk becomes rounded and her ovipositor protrudes. Then she hovers in the water and points her head upward. The male approaches and they ascend together, facing each other. At the crescendo of the process, the female inserts her ovipositor into the male's open pouch and releases up to two hundred eggs in a long sticky string. After about six seconds they separate. The eggs are inseminated in the male's pouch.

D. PREGNANT MALE SEAHORSE GIVING BIRTH. A cloud of juveniles is bursting forth from a birth pore in the pouch. Each ¼-inch juvenile can avoid predation by bypassing the dangerous planktonic stage of other fishes.

E. HOLE IN CORAL REEF showing coral polyps poisoned by tropical fish collectors. The white region around the hole and in the foreground is an area of death surrounded by living coral polyps. The dead area can have a diameter of a yard or so and coalesce with other dead areas, killing the reef.

PLATE 10

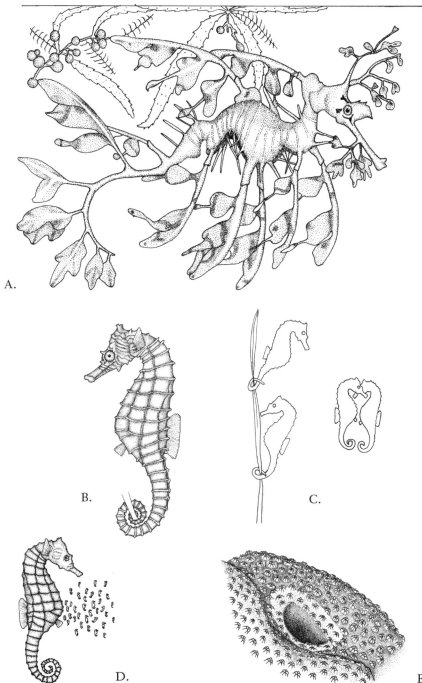

A.

B.

C.

D.

E.

What is to be done? The villagers, lured from their former lives as subsistence fishermen, have tasted the sweet sensation of money in the pocket. They have gone from merely surviving to being "rich" with earnings of a few dollars a day. They are unwilling to revert to their former poverty.

Should the government remove them from their ancestral village? The alternative for the impoverished government would be to give them subsidized housing elsewhere, impossible in developing nations. No solution has been found other than the cruelty of forced relocation without compensation.

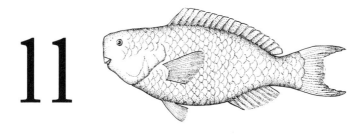

11

Super Male

DRIFTING ALONG like a gray cloud, deep in the shadows of the reef front a school of somber-colored fishes stretches as far as the eye can see. Occasionally two or three burst from the school, darting upward in a whirling pattern. At the peak of their ascent, a tiny gray cloud appears—a mass of eggs. At the moment of expulsion of the eggs, the one or two consorts augment the cloud with puffs of white sperm. The fecund mixture, an undulating diaphanous shadow, drifts off, dispersing in the currents. Another pair or triplet dart from the shadowy school and repeat the reproductive ritual. The eggs and sperm appear to be released simultaneously, but the eggs spurt out a moment before the sperm appear.

Somehow a message is received, causing the males' seminal vesicles to convulse, and they ejaculate in tandem. Evidently the presence of the eggs is the stimulus for the simultaneous release of their sperm. An invisible substance, a signal, has diffused from the egg mass. It was instantly perceived by the males, causing a mindless, automatic physical reaction. This stimulus-response sequence maximizes the chance that the eggs will be fertilized. Chemical signals are facilitated by the watery medium and are often used to coordinate physiological actions in the sea.

The spent fishes return to the protection of the school. The eggs drift off, uncared for and at the mercy of the ever-rapacious zooplankton. Suddenly, a much larger fish appears. Its exquisite colors shimmer in the filtered sunlight. It darts toward the school like a hungry predator, but the potential prey fish do not try to escape, continuing their bovine behavior, floating along aimlessly. Surprisingly, the gorgeous intruder pairs off with a female in a ritual courtship display. The preening, blunt-browed, colorful giant slants its flanks, exhibiting glowing red, green, and yellow patterns to entice the dun-colored female fish. The large fish is clumsy in its approach, and the female wanders off. Undaunted, the colorful giant begins to court another female. But again the ritual is not completed. The subject of the display does not respond in the least to the invitation of her beautiful suitor. Then the display is directed at a fish that is clearly a male. The ardent lover seems confused. First, it was unable to complete the courtship display, then it chose a partner of the wrong sex.

❧❧❧

The term describing this aggressive colorful fish is "supermale." If appearances live up to reality, that appellation seems valid. For the fish's brow has become bulbous, its tail a crescent of vivid yellow, red, and blue, its shoulders swollen with muscles. But the supermale is super in name only. Like male athletes on steroids, its appearance is not predictive of sexual performance. *The supermale is a fishy transvestite!* Recent research has revealed that the supermale possesses both sets of gonads, with incipient testicular tissue mixed in with eggs. In fact, the flashy supermale was originally a drab female! Her eggs become absorbed as the testes develop, until they are superseded by the male gonads.

The predominant mode of reproduction of these fish, parrotfish, is group spawning, but the more traditional method of paired sexual behavior is also utilized—with the supermale twist. Why evolve two sexual approaches, especially if it requires a huge amount of energy to convert a female into a male?

Using two modalities, paired sex and group sex, has an evolutionary purpose. Every once in a while the blundering supermale completes the courtship ritual, sending the right signals to a female and initiating the reproductive process. When this alternative works, it becomes evident why the survival of the species is enhanced. If circumstances prevent parrot-fishes from forming the critical mass of a reproductive-size school, this method becomes important because a reservoir of supermales can initiate paired sex.

This cross-dressing is also exhibited by the bluehead wrasse, *Thalassoma bifasciatum*. A six-inch supermale's bright bluish anterior is separated by black and white lines from the rest of its body. It is attended by an entourage of finger-size yellow females and males. Should the supermale die, one of the females will almost immediately become colorful and massive, and replace it. As with parrotfishes, reproduction is primarily by group spawning utilizing the miniature males. Only about 4 percent of the population consists of supermales, a small reservoir of traditional sexuality.

ᨃᨃᨃ Eating "Rocks": Parrotfish Predation on Corals

The reproductive process over, the parrotfishes resume their tranquil existence. As the school cruises over the bottom, one or two fishes browse on seaweed. But the parrotfishes' herbivorous origins permit an alternative food-getting style. As mixed schools of males and females wander through the channels and interstices of the ever-providing mother reef, some will take bites from a colorful coral colony. With an audible rasping sound, the parrotfish scrapes bite-size chunks from the calcareous coral head, seeking the nutritious polyps hidden in their tiny cups. The bite leaves a white gash on the pale green coral head. As it munches on its stony food, a powerful pharyngeal mill grinds up the coral chunk. Gills filter out the tiny nutritious, fleshy coral polyps. The fish defecates the ground-up coral skeleton as a cloud of fine particles destined to wash up on the shores of a tropical island as sand. (In fact, the powdery fine white sand of the tropics originates exclusively from biological sources—urchin spines, particles of coral, etc., in contrast to northern beaches that are derived from ground-down rocks.)

Eating plants requires crushing the plastic-like cellulose cell walls to get at the protoplasm locked inside. So parrotfishes are preadapted to eat almost impervious food. They can bite off pieces of coral with a parrot-like

PLATE II

A. A SCHOOL OF PARROTFISHES engaging in group spawning in front of a coral reef. A trio of parrotfish, consisting of a female and two males, rises toward the surface, releasing sperm and eggs. Males and females resemble one another and are dun colored and camouflaged, in contrast to the flamboyant supermale that practices paired spawning. The rest of school is visible in the background. Why would supermales become so flamboyantly conspicuous as to attract the attention of predators?

B. SUPERMALE BLUE PARROTFISH, *Scarus coeruleus*. Sky blue, white patch on cheek, snout distinctly humped, to 27 inches. The supermale's morphological changes are exemplified in this species by an exaggerated hump on its head. As the supermale's external physical features transition from female to male, its ovaries are gradually replaced by testicular tissue.

C. THE FEMALE BLUE PARROTFISH exhibits the original pale blue, undistinguished appearance of the supermale before its change of sex. It is advantageous for her to be dully colored so as to be inconspicuous. When group spawning, males and females are virtually indistinguishable.

D. SUPERMALE STOPLIGHT PARROTFISH, *Sparisoma viride*. Reddish fins, bright green body, yellow crescent on tail, to 20 inches. Common in the Caribbean. These and other species of parrotfishes sleep in a cocoon of mucus at night, presumably to repel predators.

E. SUPERMALE QUEEN PARROTFISH, *Scarus vetula*, feeding on coral. Green with orange highlights on fins and tail. To 24 inches. Females are gray with a wide white line on their flanks, as illustrated in *A*. The supermale is biting chunks out of coral, which pass through his muscular pharynx (a gizzard-like structure). Nutritious coral polyps are strained out and swallowed, and ground-up coral is defecated as fine grains that become sand. Tropical beaches are composed primarily of organic calcium carbonate (such as ground coral), spines of sea urchins, and dead calcareous algae, which explains their talcum powder fineness and pure white color. In background are a lavender tube sponge, *Spinosella vaginalis*, and a fan coral, *Gorgonia flabellum*.

PLATE 11

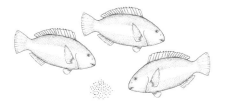

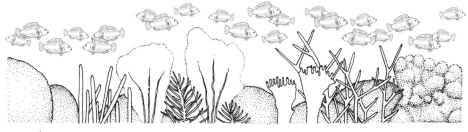

A.

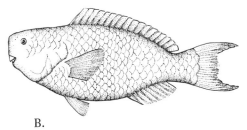

B.

C.

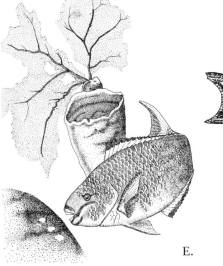

E.

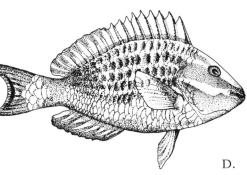

D.

beak formed from fused front teeth. These huge teeth and their herbivorous origins make parrotfishes uniquely able to feed on stony corals. They seem to have crossed the line to become carnivores. What effect does this predation have on the already-endangered corals?

A coral head is a colony consisting of clones of an original settler, expanding ever outward, following an inborn directive to create a distinctively shaped colony. Deep in the nuclei of the progenitor's primordial cells a pattern emerges. The colony becomes a pale green mound of chalky calcium carbonate crisscrossed with hills and valleys resembling the gyri and sulci of a mammalian brain. Another species with a different genetic directive produces a multiple-branched, staghorn-shaped colony. The clones, many-armed polyps, live in tiny cups, but they are attached to one another by a universal living tissue web that can heal itself. A day after the scraping bite of the parrotfish, the scar is covered by a sheet of tissue from which buds of polyps already project.

Like the Borg, the coral colony is formidable in the massive multiple effort of its interconnected clones, resisting the predation of the parrotfish. The unfettered growth resists the powerful jaws of the predator.

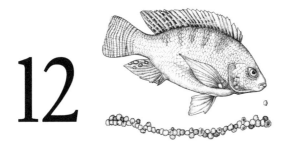

12

Miracle Fish

THE THIN, FAR-OFF STRAND of sound intruded into my awareness above the bustle of city traffic. It grew into the deafening roar of a warplane, stabbing at the kernel of fear that had lain dormant in my mind over the thirty years I had been away. The plane flew overhead, a giant blowtorch. Four tiny dots of light spurted from its tail, like a fish spawning flaming eggs. I realized these were flares to confuse heat-seeking missiles. I shuddered as I thought of the fear that must be instilled in the target by this ferocious, deafening sound—and was grateful that this was practice, and I was having breakfast on the balcony of my hotel room overlooking the port of Jaffa, in Israel. This ancient Philistine city had seen many wars, yet not since David slew Goliath nearby to the cheers of his brethren had such a din been created.

This mock air attack brought back events from thirty years before. We had arrived in Israel on a mission for the U.S. Office of Education, and settled into a nice suburb that reminded us of our hometown in the United States. But early one morning the blunt roar of a divebomber broke through the bucolic silence. Israel is such a small parcel of land that a warplane cannot properly practice its death-dance within the boundaries of the nation. Five minutes after takeoff the plane will be over a hostile country. So the pilots had taken to practicing strafing on shoreside towns. Every once in a while, an over-zealous pilot would come so close that all of us would end up butting heads under the kitchen table.

I was there to adapt modern American science curricula to the needs of Israel's schools. To my surprise, I found that the national seventh grade curriculum was tied to the culture of one species of fish, which experience had shown not even little kids can kill. I was told that this "miracle fish" is called the tilapia. It is the authentic and original miracle fish, the one with which Jesus is said to have fed the multitudes on the shores of the Sea of Galilee. It was indeed apt that children should be led in their first faltering steps of learning by this very same fish.

⋙⋙ Another Miracle

> And he . . . took the five loaves and the two
> fishes, and looking up to Heaven, He blessed
> and break, and gave the loaves . . . to the
> multitudes and they all did eat . . . and they
> that had eaten were about five thousand men,
> beside women and children.
> —ST. MATTHEW, 14:19 TO 21

The fish with which Jesus fed the multitudes was probably the Galilee tilapia, *Sarotherodon galilaeus*, or the blue tilapia, *Oreochromis aureus*. These are the most important food fish of the Sea of Galilee (Roman name: Lake Tiberias*). Most tilapia species originated in the Great Rift Valley in East Africa. Lake Tiberias is the northern-most manifestation of that huge crack in the earth. Its depths, like those of the Grand Canyon, become a receptacle for water flowing downhill. Mother of the Jordan River, it draws water from the green hills of Galilee, where Jesus tended his flocks—and supports myriad fishes, which He caught as a commercial fisherman.

The tilapia is portrayed on murals in ancient Egyptian tombs, proving that the fish has been cultivated for at least four thousand years. Their capacity for withstanding the most extreme environmental vicissitudes comes from inhabiting African Great Rift Valley rivers and lakes for countless millenia. These bodies of water were at the mercy of the cataclysmic forces that carved the deep valley. Earthquakes and volcanic eruptions

* Modern name is Lake Kinneret.

alternately dammed rivers, causing lakes to form, and, conversely, filled them with volcanic tuff and rock, making them disappear. Cataclysm after cataclysm occurred in this region where two ancient continents separate. River courses were changed, and some disappeared. Still the tilapia remained, adapting to shallow, oxygen-poor lakes, deep, cold, rushing rivers, and salty, evaporating ponds. Time elapsed. Giant lakes formed, inundating the shallow, almost dry lakes and ponds to which the tilapia had adapted. Thus, over time tilapia evolved a tolerance for low oxygen levels, high salinities, and cold water. They became super fish. It seems that however intolerable the environment, the tilapia will survive.

Lake Victoria, third-largest lake in the world, was the vessel in which three hundred tilapia-related species of fish (haplochromines) evolved in this manner. The huge lake inundated at least three small, shallow lakes. Each presented a unique environment. Each produced haplochromines and tilapias uniquely adapted to it. When their home lakes were drowned by the mother lake, they survived unchanged to form part of the lake's population today.

ꌗꌗꌗ Courting Tilapia

All tilapia of the genus *Oreochromis* are brownish with dark vertical bars. The brown conceals an underlying red pigment that becomes more or less dominant in males during courtship. Romance begins with nest-building by the male, which picks up pebbles one by one, depositing them in a shallow depression until they become an inviting signal for the female. The male then stakes out his territory, an invisible circle about one-and-a-half body lengths in diameter. Bigger males have the strength to defend larger territories, and females are attracted to these piscatorial real estate magnates. Males are much larger than females, a sure indication of territoriality. (Fish that spawn in midwater, casting their gametes to their watery fates, and thus do not need to defend a territory and lack sexual dimorphism.)

Why is the size discrepancy between males and females so apparent? In *Oreochromis* an unusual sexual phenomenon occurs. The large male performs an enticing dance, alternately approaching the female and turning away from her, undulating toward the nest. If a coquettish female needs more coaxing, he will drive her toward the nest with tiny nips. In contrast to humans it is the male who has a bag of alluring tricks. His flanks are suffused with red. This color, which seems to be associated with sex in lower vertebrates (and humans?) is particularly vibrant in some males. It has been suggested that the intensity of red is a signal to the female about the health of the male. Males with pale coloration are weaker than their gaudier brethren because it takes more energy to produce an intense red. Why? It has been shown that males producing less intense coloration are the victims of parasites. The more parasites, the feebler the color. Try as they can, pale males are too weak to produce the desirable color and must expend an inordinate amount of their energy just to stay alive. Females cruising the nesting ground are attracted to the larger, stronger, redder males, instinctively ensuring that their offspring have a better chance of survival and pushing the evolution of the species toward a healthier and larger-sized future.

᭞᭞᭞᭞ Reproductive Rituals

The male has enticed the female into his nest. The culmination of the reproductive process is about to occur. She lays a clutch of about two hundred eggs. Then she picks them up and tucks them into her expanded mouth cavity. She searches among the pebbles for the last few eggs. There are three or four more. She attempts to pick them up. But these are perfect replicas of eggs, called egg spots, that are inscribed on the anal fin of the male. The tugging on his anal fin is a signal for the male to release his sperm into the open mouth of the female. The eggs are inseminated internally, albeit in the mouth, not the vagina and fallopian tubes.

The eggs hatch in one or two weeks, having been brooded by the female in her mouth rather than in her uterus. She gives birth. The fry burst from the mother's mouth in a cloud of tiny swimming specks. Upon their expulsion they are large enough to feed and large enough to avoid the dangers of the planktonic jungle that decimates the eggs and larvae of so many fish species. The fry exhibit positive thigmotaxis, an inborn behavior causing them to press against the side of the female. When danger approaches, they seek the protection of the mother's womb-like mouth, a veritable living cloud crowding into their maternal haven. This brooding and protecting the young mimics infant rearing by humans. The protection of the juveniles explains the ability of the species to overpopulate lakes in the same manner that humans are overpopulating the earth.

Some states prohibit the introduction of tilapias. Florida has an adequately warm temperature for these African fish. They were introduced in the 1960s as a potential food fish. Now a less-than-valuable species of tilapia can be found in every drainage ditch in the southern part of the state. Soon they will enter Lake Okeechobee, threatening extinction of its indigenous fish populations.

Why do tilapia populations not dominate Lake Victoria, whence several species originated? Because predatory species evolved along with more fecund species. For example, four haplochromine species of the genus *Cyrtocara* are pedophages (egg eaters). Some species approach a pregnant female and ram her, forcing her to expel the eggs, which become a high-energy food for the predator. Other species just suck the eggs from the female's mouth. In another highly specialized mode, an unusual predatory species cuts circular holes in the flanks of prey fishes, similar to the style of cookie-cutter sharks.

Why does the female tilapia produce only a few hundred eggs, in contrast to other species that can produce millions of eggs? Only relatively few of these large, yolky eggs will fit into her mouth to be brooded. But brooding, as we have shown, insures the safety of the young past the dangerous larval stage, so more survive.

In the early 1950s in Taiwan, a few mutant tilapia were born with an inability to produce the natural brown pigment. The underlying red pigment became dominant. They were bred to become the progenitors of a red-tilapia-based industry. In countries like Jamaica, farmed normal-colored tilapia were introduced as "African perch." They bombed. It seems that the Jamaicans were used to eating colorful coral reef fish and these brown fish did not appeal to them. But an entrepreneurial fish farmer imported some of the red tilapia. Presto! People liked the "Cherry Snapper" or "Jamaican Red" tilapia. Today, the culture of tilapia has become highly sophisticated. Color variations have been bred to suit the tastes of a variety of consumers—red, white, blue, and gold. Tilapia farming has become the fourth-largest aquaculture industry in the United States.

❧❧❧ The Tilapia That Saved My Life

It all began with a phone call. The man at the other end asked if I knew about tilapia. The caller was a research physician, part of a team that specializes in heart transplants. But this man was a specialist in diabetes, not heart disease. He had heard that tilapias have an insulin-producing structure that is composed only of insulin-secreting cells. The human source of insulin, the pancreas, is composed of several kinds of cells. He wanted to transplant the tilapia insulin-producing tissue into human diabetics. "Why not a human pancreas?" I asked.

He answered, "Because the human insulin secretory cells are embedded in pancreatic tissue and, instead of having to fight the immune system with several kinds of cells, these fish can provide pure insulin-producing tissue."

"Why not another fish?" I queried.

"Because the tilapia is a tropical fish and its organs can survive at the high temperatures inside the human body." I provided him with a supply of living tilapia and an eager young graduate student to work on the tilapia end of the project.

What goes around comes around.

Many years later I was diagnosed with a brain tumor. What surgeon should we pick to perform this delicate operation? After several unsatisfactory interviews, I thought of my friend, the tilapia transplant researcher. He recommended a former prize student who had become a neurosurgeon. The man performed an exemplary operation, removing a tumor the size of a baseball from my head. End of story.

❧❧❧ Tilapia Farming in Africa

Scene shifts to another year and another place—my lab in Jamaica. At lunch one day a huge, prepossessing man sat opposite me, looking myopically at me through thick glasses. We chatted. He was a vice president of a major brokerage firm and had established a fish-meal plant in Australia in his youth. He was a fellow fish-person. A year later, I received a call from him. "Do you want to help establish a tilapia farm in Africa?" he asked.

Eventually a team was put together: my financier friend, myself—in charge of the design of the farm—and a bilingual businessman from the West African nation of Côte d'Ivoire (Ivory Coast). This country became the financial powerhouse of West Africa under its founding president, who developed a style of governance similar to the traditional chieftancy. In ancient West African tradition, the ruler remained in power by dispensing patronage to the leaders of the individual tribes. This president modernized the technique, doling out superhighways and housing projects using the proceeds of taxes on cocoa and coffee, thus keeping the peace. The population of Côte d'Ivoire is composed of more than twenty tribes, none of which is dominant, which for a time provided stability and avoided the tribalism rampant in much of Africa.* The financial capital, Abidjan, is a bustling modern city of skyscrapers and a six-lane superhighway. The skyline is dominated by an exquisite Catholic church graced with a huge cross whose curved, enveloping arms seem to invite entry. A nearby gigantic mosque signifies the ecumenical approach of the government. The United States is represented by a nearby slum, called "Washington DC."

Our Ivoirian colleague was president of a prosperous construction firm. Upon visiting his offices, we were struck by the friendliness and courtesy with which each member of the staff treated one another. These were nice people. Computers hummed, blueprints were produced, building projects were developed. We met our potential partner's wife and mistress, and, in that homey environment, decided to build the farm.

Tilapia are not native to West Africa. To our surprise a government agency had already chosen a species of tilapia from Egypt for its rapid growth rate and resistance to environmental extremes. For the site of the farm I chose a lake only half an hour away from Abidjan by dirt road and highway. Our fish would arrive at the Abidjan markets so fresh as to be

---

* Sadly, upon the founding president's death, politics in Côte d'Ivoire created a divisiveness where none had existed. By political manipulation and the evil legacy of war in neighboring countries, the nation has been riven into a north-south rivalry, with attendant destruction of a vibrant economy and the slaughter of innocents.

PLATE 12

A. MALE NILE TILAPIA, *Oreochromis niloticus*, showing egg spots. Wide brown stripes on tan background, to 14 inches. Bred commercially to produce pinkish red, gold, and white colors. The male is building a nest. He will attempt to entice a female into the nest by flaunting his red-tinged fins and dorsal surface.

B. A FEMALE has spawned on the nest, releasing more than two hundred eggs. The eggs in the nest and a male's anal fin are shown. The female will pick them up and retain them in her mouth, where they will be fertilized. With her mouth full of eggs, she will tug at the egg spots on the male's anal fin. That will be the signal for him to ejaculate into her mouth.

C. JUVENILES instinctively pressing against the flank of the female. At the approach of a predator, fingerlings stream into her mouth for protection. After about three days juveniles will leave the mother's protection.

D. GIVING BIRTH. The female releases 200–250 juveniles, each about ⅜ inch long. They have spent about two weeks in her mouth undergoing maturation and are fully formed and ready to feed. The brooding process insures that juveniles will not have to go through the dangerous planktonic phase of most fish development.

E. A PEDOPHAGE (egg eater) engulfs the mouth of a female to prey on her eggs. This is an example of adaptation among fish species closely related to tilapias, the haplochromines. These species have invaded virtually every kind of habitat in Lake Victoria, and have physically adapted to virtually all of them. Some haplochromines live near the surface and are silvery; others live near the bottom and are brown. Some are predators on smaller fishes, others are predators of eggs. The strategy used by this species is to engulf the mouth of a female and force her to expel her eggs, which it then swallows.

Tilapias, like the haplochromines, exhibit diverse adaptations they have made over thousands of years and can live in most aquatic habitats, from seawater to poorly oxygenated muddy ponds.

PLATE 12

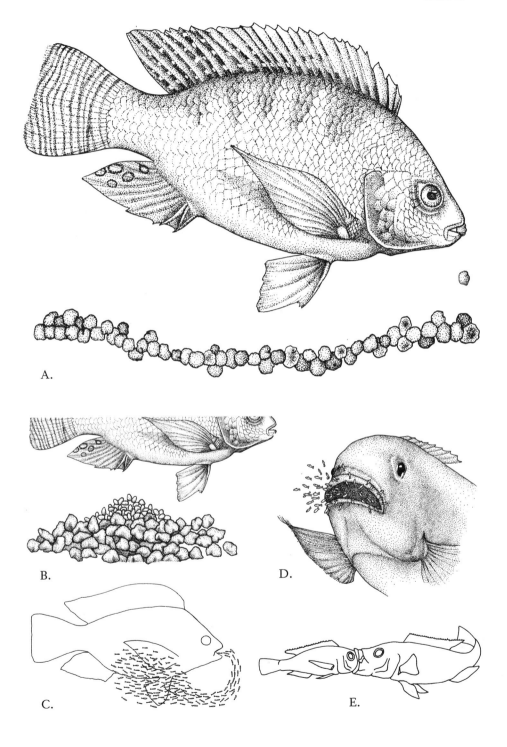

A.

B.

C.

D.

E.

virtually alive after a half-hour journey in insulated fish boxes, in marked contrast to the dried-up, fly-covered tilapia shipped from two hundred miles away. Years passed. Slowly the farm grew from a man-made clearing in the bush to a series of small buildings comprising offices, a laboratory, storage sheds, and a set of huge concrete breeding tanks designed to provide juvenile tilapia for growout in gigantic mesh cages set out in the lake.

Everything was finished. Then the sword fell. The Ivoirian partner became bankrupt. Sadly, my friend, the founder of the project, died. I was left with a farm that had produced 250,000 juvenile fish and no money for their food. Disaster!

Not a businessman, I had sought not money, but to develop a model for inexpensive protein production in developing nations. The most important indigenous food sources in Côte d'Ivoire were an expensive catfish (which is the national symbol), chickens, and a rodent called an agouti. Despite the near-ruinous events, my hope was realized! After a year or so our company was bought out by an Ivoirian firm and the farm is operational to this day.

I replied, smiling my most convincing smile. She held it up, staring at the stringy thing, uncertain. Her analytical examination clearly switched its focus. "Why is *he* staring at us?" I looked in the direction of her glance. There in the corner was a man in an apron, clearly the chef. There was no doubt he was staring fixedly at us. My voice lacked conviction when I said, "There you go, being paranoid again." I looked back. He had disappeared.

"This place is getting creepy," she said. The waiter brought tea. She began to relax—the night's saki making her brave. "Why isn't there a menu?" she asked. We were surprised when the waiter placed lovely platters in front of us, clearly the main course. The food was beautifully arranged in an intricate floral pattern. "Why do you get more than me?" The booze was bringing out the beast in my otherwise discrete friend.

"I'll give you some of mine," I answered, "if you like the stuff." "Oh I'll like it, it's so pretty—I hate to spoil this nice design." It tasted mildly fishy and a little sweet. It was clearly raw fish. "This is good, like sushi without the roll," she said. We ate with gusto. Our diet of carbs in the form of noodle soup left us ravenously hungry. This was a fitting end to the night's splurge.

"I feel funny," she said. My lips were beginning to tingle—a pleasant sensation. I smiled. "So do I." "I'm getting dizzy now," she laughed. The room began to whirl, and I tried to stand—but fell back in my seat muttering, "You're drunk—eat your dinner and it will straighten you out."

"That's funny, the cold fish tastes hot," I thought. I looked at the remnants of the beautiful fish flower on my plate. Their shape was grotesque. What had been neat little petal-like morsels looked like big, jagged geometric shapes.

I pushed off from the table. My fingertips felt numb. Her brow was sweaty, and her eyes began to glaze. I said, "Let's get out of here." She tried mightily to stand. I felt weak, but I grabbed her, left I don't know how many yen on the table, and, clutching each other tightly, we staggered off into the night.

13

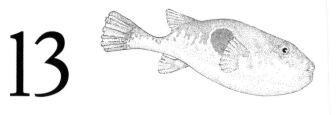

Fugu

"A man who is his own sushi chef has a fool for
a customer."

—JAPANESE PROVERB

WITH A LADYLIKE LITTLE STAGGER, she said, "Let's get out of
lights are getting to me." The hyper nightlife of Tokyo's Ginza w
and the screaming two-story-high neon lights flashing glaring r
prehensible messages were giving me a headache.

"I need seafood," she slurred, "There's a place." In the dimness
alley appeared a small wood oval sign. It was carved in the shape
With my arm draped over her shoulder, we tipsily entered the din
ingly quiet restaurant. I noticed that the only clientele seemed to l
Japanese men. "Why are they staring at us?" she said. "You're j
paranoid," I replied. But everyone in the small restaurant seem
looking at us, each with a mildly amused smile.

We sat at an intimate round table and a waiter brought chops
some indistinguishable appetizers. "What's this?" she said, gigglin
held up one of the less-appealing items lying flabbily in a small
appeared to be a squid tentacle, replete with suckers. "It's only a r

This encounter with ancient Japanese culture ends well. But if the chef had miscalculated, the story might not have had such a happy ending.* We had inadvertently wandered into one of several "Fugu" restaurants in Tokyo.

Fugu, genus *Takifugu*, are species of puffer fish. In the wild or on ice in the market, they look cute. Chubby, with eyes placed forward to give it binocular vision, the face seems human—or if not quite human, then uncannily like the face of a bulldog. Virtually all puffers are able to imbibe huge quantities of water. A valve prevents expulsion and when stressed, the fish can swell its body until it becomes distended to two or three times its normal size. When a puffer is brought to the surface by a fisherman, it takes in air instead of water, giving rise to its name, "blowfish." Bad Little Boys (see chapter 21) pick up the living balloons and toss them to one another. Invariably, the air is expelled in mid-flight and the fish drops like a stone.

One family of puffers is covered with very sharp spines, making them indeed formidable when inflated. Imagine a predator facing this intimidating array of defensive weapons. First the puffer becomes huge, too big to attack, then half-inch spines bring the message home, "Stay away." Another family of puffers produces a very powerful poison, tetrodotoxin, one of the most effective neurotoxins known to man. One wonders why this almost-invulnerable balloon of a fish needs this powerful poison in its tissues. Added to its unpleasantness is a bad smell emanating from the puffer's flesh when it is cleaned. It also is a vicious feeder when in the wild. While most puffers use a beaklike fusion of the front teeth to feed almost exclusively on snails, some tropical species are opportunistic feeders—and adding to their gruesomeness, have been known to snap at the testicles of swimmers.

We brought back from Jamaica a porcupinefish, *Diodon hystrix*, and raised it in a large aquarium in the classroom. It was very popular because it was considered "cute," due to its aforementioned resemblance to a miniature bulldog. There were other fish in the tank, and there was much competition when we appeared once a day to drop in the day's meal at the customary corner of the tank. A feeding frenzy ensued. All the fish in the tank rushed to the feeding corner and the water was churned into a froth.

---

* The effectiveness of poison is a function of body weight. The chef looked carefully at the couple to estimate their weights. That is why the young lady was served a smaller portion than the hero. She weighed less.

The puffer was the biggest fish in the tank, and it became trained to lunge at the food the moment it hit the surface. If a smaller fish was in the way, it was chopped in half. One day I came upon some of the Bad Big Boys in the class trying to tease the fish. They dipped a pencil in the water at the feeding corner. The puffer instantly darted toward the "food"—and chopped the pencil in half! We worried about the possibility of someone putting his pinky into the water, but it never happened.

The custom in Japan is to eat fishes of the genus *Takifugu*, commonly known as Komonfugu in Japan, Tinga Tinga in the Philippines, and Bok Oh in Korea. It is served in restaurants devoted to this genus only and is prepared by chefs trained in this specialty. The cookbook they use is the only book in the world devoted to cooking poisonous animals. The licensed chef examines the customer, guesses his weight, determines his or her sex, and estimates the potential resistance to the toxin. If he is wrong, the patron gets more than he bargains for. If he is right, the patron reaches the edge of death and the desired sensation: lips tingling, vision blurry—all the symptoms of intoxication but with the added fillip of potential death to add excitement.

Miscalculation by the chef can bring about a hideous death. It seems that in some cases the total paralysis that results simulates death but can leave the victim alert and aware. There were occasions when a paralyzed victim was thought dead and was buried.

Why do the Japanese do it? I don't know. Bungee jumping seems to be a preferable hobby.

Eating fugu is probably an ancient custom performed for centuries by jaded elderly gentlemen. Naturally, the Japanese are the leaders in legislation and research into the toxicity of puffers. The minutes of the Tokyo Health Department read, "The Tokyo Prefecture has instituted one of the better health programs regarding the sale of fugu in Japan. The law was termed The Professional Globefish Treatment Regulation . . . the law was poorly enforced and . . . new laws were passed by the Tokyo Legislature as Regulation No. 43, Regulations for Handling Fugu, dated April 5, 1949 . . ."* As a consequence, all cooks and restaurants handling fugu must be licensed.

Every species of puffer has some degree of toxicity. A delicacy on the East Coast of the United States is "sea squab," locally called the blowfish,

---

* Bruce Halstead, *Poisonous and Venomous Animals of the World* (Princeton, NJ: Darwin Press, 1978), 464.

Sphoereoides maculatus. It is prepared by removing the smooth skin with one incision and ripping out the muscular abdomen of the fish, excluding the entrails. The result is a tempting-looking frying-pan size morsel that looks much like a drumstick. Luckily it is only the abdomen that is eaten. The liver is toxic, but if a little liver is left on the piece of fish you are eating, don't worry. The toxin is destroyed by cooking.

Ciguatera

One version of toxicity in fishes occurs in the Caribbean, and its distribution includes many species of fishes and many places. Occasionally people are paralyzed or die after eating fish. The death and paralysis are not attributable to any species of fish or any area of the Caribbean. Confusion reigned concerning the nature of the condition and the causative agent up until the 1970s, when intimations of what was happening began to appear. The ephemeral nature of the condition presented an almost insoluble mystery. There was nothing to hold on to. When people ate fishes that were known carriers, they didn't get sick. When people ate those same species on another day or on another island, they came down with the condition. Despite its vagueness, the condition existed. To focus on it, a name was needed. The causative agent was called ciguatoxin and the condition was called ciguatera.

The mystery led to cultural preventive methods:

- If you drop a dime on the suspected tissues, it turns black if the fish is a carrier.
- Look down the throat of the fish. If it appears striped orange, it is ciguatoxic.
- Put a piece of the fish near an ant hill. If the ants refuse to eat it, you should refuse, too.
- Put a bit of flesh in the sun to dry. Eating the fish will kill you if flies refuse to settle on it.

The first clue was the fact that all of the carriers are predators, with barracuda flesh being one of the most frequently associated agents. But other predatory fishes such as mackerels, jacks, and groupers were implicated.

PLATE 13

A. FUGU, *Takifugu rubripes*. Blackish-brown above, white below, one or several large black spots. To 20 inches. Ovary and liver are strongly toxic, intestines and skin moderately toxic. The price in Japan can reach $50 a pound, and sashimi can cost $500 a plate. A chef-in-training must take an exam: prepare paper-thin slices in twenty minutes, labeling toxic parts with red tags and edible parts with black tags. Of 900 applicants in 2004, 63 percent passed the course. Eating liver has been banned in Japan since 1983. Before that, hundreds died.*

B. FUGU-CHIRI are pieces of fugu boiled at the table with *konbu*, a seaweed. Various vegetables are served with the fugu. *Fugusashi* (not shown) are translucent strips of fish as thin as leaves arranged around the fish's skin. They are served with the fin, or *hire*. The fugusashi can be dipped into *shouyou* (soup containing ovaries and other organs).**

C. CARIBBEAN PORCUPINEFISH, *Diodon hystrix*. Tan with brown spots, to 35 inches, erectile spines to ¾ of an inch. Laws against the sale of this species have been in existence since 1884. Liver, ovary, and skin are toxic, but not the muscles.

D. COMMON EAST COAST BLOWFISH, *Sphereoides maculatus*. Light brown above, white below, to 12 inches. Liver and ovaries are toxic. The nontoxic, drumstick-shaped muscular portion of body is commonly eaten as "sea squab."

E. A MICROSCOPIC DINOFLAGELLATE. A single cell with one vertical and one horizontal groove, each containing a flagellum. The central sphere is the nucleus. Irregular dark structures can be chloroplasts or other pigmented bodies. Oil globules are also visible. One species, *Gambierdiscus toxicus*, has been implicated as the source of ciguatoxin. Other species cause red tides and other toxic events.

F. GREAT BARRACUDA. Green to gray above, silvery flanks, black blotches, to 5 feet. A common cause of ciguatera when the flesh is eaten.

---

* L. Kaufman, "Why are Coral Reefs so Colorful," *National Geographic*, May 2005, 86.

** Dish descriptions from Nick May, "Eating Fugu," *Fukuoka City Guide*, 2005, http://kyushu.com/fukuoka/features/fugu_1.

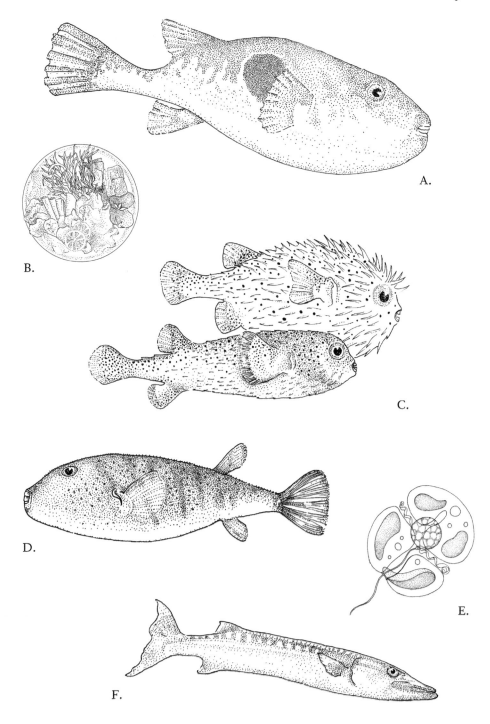

PLATE 13

A.

B.

C.

D.

E.

F.

While barracudas are eaten primarily by locals in the Caribbean, groupers are popular in hotels. When an islander gets ciguatera, he just stays in bed for up to six months. When a tourist gets ciguatera, a big fuss is made. The symptoms of this neurotoxin are similar to that of puffer poisoning. First the lips tingle, then the fingertips become numb. Hot feels like cold and vice versa. Breathing becomes difficult. In long-term cases, clinical depression can occur. While the condition can be treated symptomatically (each aspect dealt with individually), ciguatoxin poisoning can be serious if untreated. A few people die each year, mostly individuals who do not have access to physicians.

The next puzzle, the fact that its distribution is spotty, led to a rash of hypotheses based on some localized etiological agent. The theory was that a predatory fish ate many prey fish and the toxin was in the flesh of the prey. Its concentration in the body of the predator became higher and higher with each toxic fish eaten; this is called biological amplification. But why was one predatory fish ciguatoxic and another, from the same area, not? Is it the predator's prey preference? Does one barracuda "like" silversides and another "like" anchovies? Not likely.

The best answer has been contributed by Japanese toxicologists. They have the most experience with fish-borne toxins. Here is their thinking: Many species of herbivorous fishes graze on bottom-dwelling seaweed. The seaweed often wears a living coat comprised of many species of bacteria and protozoa, including toxic organisms called dinoflagellates. Most dinoflagellates are harmless, but a few species produce toxins. In this instance the culprit is aptly named *Gambierdiscus toxicus*. When the herbivore eats the algae, it picks up the toxic dinoflagellates, the poison of which accumulates in its tissues. When a predator eats the herbivore, it is, in effect, eating a poison pill. The toxin in its tissues doesn't seem to affect the predatory fish, but its random feeding on various herbivores may make it ciguatoxic. The deadly dinoflagellate populations wax and wane seasonally; they are found in specific locations depending on currents and light levels. That may account for the randomness of the distribution of ciguatera.

Don't worry about the fish you eat on your next vacation. Ciguatera is very rare. If you are a thrill seeker, you can try fugu for a delicious taste of near-death. Eating raw barracuda is a long shot at best.

14

Bunnies of the Sea

THE NIGHT IS DARK. Loaf-shaped shadows move in the shallows. Olive drab, camouflaged with black leopard spots, they sluggishly slime their way through the sparse sea grass. A flashlight's beam reveals giant shell-less snails, some almost a foot long, moving at a literal snail's pace toward the darkness outside the light's halo. The head is adorned with two pairs of projections. The first, flat like the ears of a rabbit; the second, antenna-like. It is the first pair that identifies these huge snails as *Aplysia dactylomela*, the sea hare. And bunny-like they are.

Sea hares mate frequently. They find each other by smelling the water with their antenna-like olfactory organs called rhinophores (rhino = nose). Often the sensual aroma of sexual pheromones so permeates the water that many sea hares are simultaneously stimulated and a veritable orgy ensues. A pair couples. The male inserts his penis into the gonopore of the female. But wait! The sexy smell has attracted another sea hare, that approaches and aligns itself behind the first "male." It inserts its penis—into the vagina of the "male," which has just impregnated the "female" ahead of it. Only one explanation is possible. Sea hares are *hermaphrodites*, having two sexes inside one body. The pheromones have aroused other sea hares. These move slowly, inexorably, toward the first trio. They line up, each acting as a female to the one behind it and a male to the one ahead of it—a sexual parade! But this parade is essentially unfair. Why? Because the first sea hare can function only as a female. Being at the head of the line is, in this instance, to be deprived—deprived of fully participating in the mating ritual and acting as a male. But evolution has resolved the problem. If you are as smart as a sea hare, you will be able to resolve this problem as they do. (If you are not as smart, check out the footnote.*)

When on a field trip, we often come upon sea hares. When we do, an ancient (twenty-year-old) ritual is reenacted. A sea hare is passed from hand to hand among the students. Some shriek at the Jello-like consistency, others almost drop the animal because of its substantial weight. But everyone performs the ritual. The animal is caressed or lightly squeezed. At that stimulus it secretes a viscous purple fluid. The tradition is to press your purple-covered hand against your white T-shirt, forever implanting the symbol of the Fraternal Order of *Aplysia* Lovers (FOAL). Whether in Rome or London, Pittsburgh or Los Angeles, members are sworn to hug anyone with a purple handprint on his or her T-shirt.

The purple secretion, common among snails, is thought to be a defense mechanism. When attacked, the mucous-laden dye is released to undulate on the currents, becoming a distraction to the predator, drawing it away and allowing the sea hare to hide. My latest million-dollar idea is to establish a sea-hare farm. I visualize "milkmaids" squirting purple dye into tiny "buckets." The collected dye would be of value to a cosmetics company that would pay many dollars per ounce for the purple fluid. But the dye is mixed with mucus, which may cause it to lose its appeal as a cosmetic. If that fact becomes known to the glamorous woman who is delicately applying it to her eyelid, my get-rich-quick scheme will be thwarted. Don't tell.

꜆꜆꜆ Super Snails

There are traditional snails that crawl sluggishly on the muddy bottom, eating rotting organic matter. And there are the super snails, the opisthobranchs. They represent an extraordinary efflorescence of the gastropods (= stomach-footed animals).

A sea hare is no more like an ordinary snail than an eagle is to a sparrow.

---

* They make a circle.

First, its shell has been reduced to small, chalky chips inside a hollow chamber surrounded by a muscular wall, the mantle. By this reduction of the cumbersome shell, it becomes the speedster of the snail world, moving at a super-snail's pace. Second, the mantle cavity has evolved from an internal space to an external groove running longitudinally along the back. In conventional snails the gills are hidden deep inside a buried chamber, like lungs inside a terrestrial animal's body. In super snails the gills are exposed. Sea hares have an open "gill chamber," the walls of which flap back and forth, forcing oxygenated water over the gills as a fish does. The multifaceted mantle cavity contains the penis and the openings of the urethra, the anus, and the female gonopore. By raising and lowering the flap-like mantle walls, the huge snail can swim slowly and gracefully.

One species of sea hare flamboyantly displays long, flowing leafy extensions that project from its back, identical to the forked-frond seaweed on which it lives. Hordes of them were invisible to me until I noticed a clump of fronds not moving with the ebb and flow of the waves, but pulsating with a rhythm of its own. I touched the fronds. They shrank away from my hand in unison. Suspicious, I combed the seaweed with my fingers, and came up with a ragged sea hare, *Bursatella leachii pleii*. Inch-long papillae projecting from its back, exactly resembled the brownish forked fronds of its seaweed haven.

A group of neuroanatomists at Columbia University is studying the sea hare's nervous system. Because this animal has a simple nervous system with large neurons, it is possible to trace the complete neuronal pathway from the stimulus to the response. Just as the fruit fly was fortuitously chosen to contribute to our understanding of genetics, and the squid's giant nerve cell axons became the classical subject for investigations of the nerve impulse, so has the noble sea hare become the palette from which the fundamentals of animal behavior have been painted.

~~~~ Unscrupulous Farmer

One often encounters a juicy-looking green slug-like animal atop a blade of turtle grass, flaunting its white fringed, ribbon-bedecked back for all to see.

This lettuce sea slug, *Tridachia crispata*,* appears to be easy prey to passing fish. Why has its evolutionary imperative taken it from the dark protective shadows into the sunlight? The super snails excel at one particular skill—thievery. The lettuce sea slug eats algae. It pierces the algal cell with a stylet and sucks out its juices. These juices are speckled with green dots, chloroplasts, microscopic sacs of chlorophyll. The tender sacs pass through the gut of the slug, *remaining intact*, and into frills extending from the back. The reason that it must perch at the pinnacle of grass blades out in the open—a death-defying habitat—is that the sea slug obtains a substantial portion of its energy budget from photosynthesis performed by chlorophyll stolen from the algae. It farms the chloroplasts on its back! Thus it cannot hide and must disport itself in the open. In theory, if a lettuce sea slug were to be deprived of its algal food, it would survive as long as it is bathed in sunlight.

Eons ago sea hares and sea slugs, exploring their evolutionary options, "chose" to feed on plants rather than animals. But plants are a poor energy source. Much grazing is required to acquire the amount of energy obtained from animal prey. The huge sea hare spends virtually all of its time feeding. It is the aquatic equivalent of an elephant. The lettuce sea slug goes one step further than grazing. It becomes part plant. But here we face another evolutionary dilemma. In order to bathe its chlorophyll-laden frills in the sun, the lettuce sea slug must leave the refuge of dark shadows and flaunt itself in the sunlight. Every fish that comes by would find it a delicious morsel. What's the gimmick?

A fish approaches. It sees the sea slug and in a moment the slug is sucked from its perch and into the mouth of the fish. But after a few seconds the fish spits it out, alive. If a fish could exhibit emotions, its face would be screwed up in revulsion. In giving up its traditional snail shell, the slug has had to evolve another protective device. The skin of the attacked slug exudes a gray secretion from scattered glands. Their name has evolved from the same source as a common expression used in the dating game. After a

---

* There has been some sentiment to replace the genus *Tridachia* with *Elysia*, but I suggest that the organism deserves its own genus. Its morphological resemblance to the other members of the family Elysiidae is superficial, and it would be misleading to lump it in with other members of the family.

first date with a person whose breath is redolent of garlic and onions, a friend asks about the experience. You wince, saying, "He(she) was repugnant." The slug is as repugnant to the fish as your date was to you. Its skin is covered with repugnatorial glands. After a few tries, fishes learn not to eat the lettuce sea slug.

ᔧᔧᔧᔧ Naked-Gilled Carnivores

A relative of the plant-eating sea slugs, the carnivorous nudibranch (= naked gills) is possibly even more remarkable than the sea slugs in its variety of colors and shapes. The Spanish dancer *Hexabranchus sanguineus*, found off the Great Barrier Reef, is almost as large as a sea hare. It flashes scarlet and black as it sinuously "dances" through the water. The leather doris nudibranch, *Platydoris angustipes*, a blob of orange tissue with a tuft of feathery frills coming from its back, demonstrates naked gills. "Fingers" project from the backs of many nudibranch species. These thin-walled projections, cerata, often contain extensions of the gut. If a nudibranch is eating its way across a red sponge, the contents of its gut are visible in the cerata, perfectly camouflaging the animal against its sponge background.

Nudibranchs have evolved to be carnivores, in contrast to the herbivory of the sea slugs. They, too, graze. But their meadow is likely to be a colony of cnidarian polyps. Virtually invisible as it undulates across the colony, its gut and cerata bulging with polyp prey, the nudibranch devours all polyps in its path, leaving a trail of death behind it. But the polyps have a formidable defense mechanism: the devastating nematocysts. The catch-22 is that to attack the polyp colony is to invite a shower of poison darts, and to avoid the colony is to starve to death. How do the nudibranchs survive this hail of poison-tipped arrows? The answer is that the nudibranchs have evolved a technique to resist the effects of the nematocysts. Are the nudibranchs born with some mysterious mechanism to resist the poison, like an armored body? No. Researchers could not uncover the secret until someone observed a species of nudibranch in the act of protecting itself from the nematocysts. It approached the colony of polyps and touched the margin. It

PLATE 14

A. Spotted sea hare, *Aplysia dactylomela*. Olive with black donut-shaped spots, to 16 inches. A pair of mantle walls surround a chamber that contains gills, siphon, reproductive openings, and a club-shaped calcareous sliver, its vestigial shell. Lacking the typical heavy shell of conventional snails, some species can swim by flapping mantle walls. This sea hare has emitted a dark purple dye mixed with mucus that is forming confusing swirls to distract a predator.

B. Lettuce sea slug, *Tridachia crispata*, exposing its chlorophyll-filled cerata to sunlight from the top of a turtle grass leaf. It has obtained chlorophyll by using a stylet to pierce algae, sucking up chloroplasts. Gut extends into cerata. Usually the upper surface appears to be covered by greenish, white-topped frills. Sea slugs can be differentiated from nudibranchs by their choice of food. Unlike the carnivorous nudibranchs, they eat algae.

C. Dorid nudibranch. The tuft of unprotected gills projecting from its back demonstrates why members of this group of mollusks are called nudibranchs (naked gilled).

D. Three *Glaucus atlanticus* nudibranchs attacking a jellyfish, *Porpita porpita*. These pelagic nudibranchs are unusual in that their cerata have evolved into floatation devices using flattened finger-like projections. They feed exclusively on jellyfish. Most other nudibranchs eat bottom-dwelling polyps.

E. Neapolitan nudibranch, *Spurilla neopolitana*, grazing on a colony of polyps and leaving a bare white trail. Its cerata contain nematocysts stolen from the polyps. Many nudibranch species live on one species of polyp all their lives, crisscrossing the colonies with trails of death. The polyps soon regenerate, repopulating empty areas. Several nudibranchs attacking a colony will kill it. *Spurilla* eats many kinds of polyps, including sea anemones, and has been known to differentiate between orange and white polyps of the same species of cnidarian.

F. Mating sea hares forming a characteristic circle, each one acting as a male to the one in front of it and a female to the one behind it. Each possesses a penis and a female gonopore.

PLATE 14

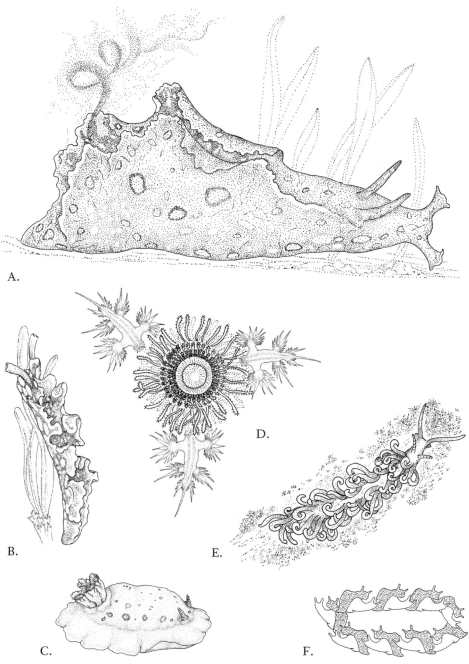

A.

B.

C.

D.

E.

F.

reared back like a horse. Again it attempted to crawl onto the colony; again it recoiled. After many tries, it was able to crawl onto the colony. Apparently the repeated bombardment by the nematocysts had caused it to produce copious amounts of mucus. This was its protective garment.

Camouflage is often a poor defense against sharp-eyed predators. Here is where the idiosyncratic skill of the naked-gilled carnivore is called upon. The nudibranch eats the polyps, gaining nourishment from their tissues. The polyps' nematocysts, in their protective cells, are inconsequential as a source of food. But they can have another role. They are stored in the cerata projecting from the back. When a predator mouths the animal, it is showered with the pilfered poisonous projectiles. This is all the more impressive when one realizes that the cells containing the nematocysts literally have a hair trigger. To touch one is to discharge it and be poisoned. Amazingly, these super snails have evolved a mechanism to pass the nematocyst-bearing cells through the gut wall and into the cerata *without discharging them.* The nudibranch survives to wander over the polyp colony, leaving a bare, meandering trail of devoured polyps. A few days later the trail is covered over with newly cloned polyps. All is well with the colony and its predator.

One of the most aberrant nudibranchs, *Glaucus,* is a wanderer. Forsaking the life of his bottom-bound fellows and taking wing, he swims through the water near the surface. Four arm-like appendages have sprung from his body, each tipped with flat finger-like extensions, giving this pelagic nudibranch the surface area it needs to float and a means of mobility. A ferocious predator, it hunts down jellyfish. A pack of hungry *Glaucus* surrounds a hapless jellyfish like wolves attacking a stag, devouring it until it is no more.

Glaucus was named after a mortal-turned-god. Like his namesake, he was a fisherman. One enchanted day he watched his silvery catch writhe out of his net as if possessed by some magic, escaping back into the sea. He plunged in after them and began to drown. The sea gods, Oceanus and Tethys, took pity on him and turned him into the original merman. "His hair was sea-green . . . and what had been his thighs and legs assumed the form of a fish's tail."*

* *Bullfinch's Mythology, The Age of Fable* (London: Hamlyn, 1964).

One day Glaucus saw a fair maiden named Scylla and fell in love with her. She spurned him. In his misery, he turned to the enchantress Circe, asking her for a love potion. Circe took a liking to the handsome Glaucus and tried to seduce him—to no avail. Unable to punish a fellow god, she took out her anger on poor Scylla, turning her into a six-headed monster. Each head was crowned with serpents and had fierce jaws. Her change from a lovely maiden to a monster upset her and she took to biting off the heads of passing sailors.

This melodrama plays itself out in the ocean: poor Scylla has become *Scyllaea*, the Sargassum nudibranch. It lives on a species of golden-brown *Sargassum* seaweed that floats perpetually on its air-filled bladders, never touching the bottom. *Scyllaea* is perfectly camouflaged, bearing golden frond-like extensions of its cerata. Like her namesake, she bites off the tops of jellyfish polyps.

15

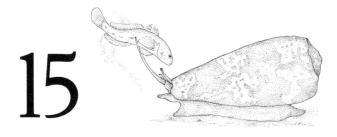

Passion for Purple

ONCE UPON A TIME (ca. 1200 B.C.) a man was wandering along the seashore. A crunching sound underfoot led him to examine the sole of his sandal. Crushed into a smeared mass was an inch-long snail surrounded by a halo of purple. History was changed at this moment and destiny dictated the rise and fall of empires based on the purple secretion of this abundant Mediterranean rock snail, *Thais haemastoma*.* This man, his identity lost in the clouds of time, scratched his head and thought. His musings created an industry that made his city-state Tyre, famous, and his nation, Phoenicia, (now Lebanon) a major trading power. All of this was based on purple dye extracted from the snail.

The dye was called Tyrrian purple, after the city that held a monopoly on its production, or "royal purple," after its use. Its production was cleverly restricted by the Phoenician entrepreneurs so that it became so valuable that only kings could afford to purchase it on a large scale. Sales of this product were such that the Phoenicians became sailor-traders, and their vessels sailed to the farthest reaches of the then-known world, the Mediterranean basin.

Royal purple became the symbol of freedom—the ability to determine

---

* Another duo of snail species whose purple secretion provided the dye for Phoenicians and Amerindians may have been *Murex trunculus* (or *M. brandaris*) (Mediterranean) and *Purpura patula* (Caribbean). According to one reference, eight thousand snails were required to produce a thimbleful of purple dye, explaining why it was so precious that it was affordable only by royalty.

one's fate, an unusual status for individuals under the oppressive rule of conquerors and kings. This color-symbol is found in the prayer shawl, the talit, of Jews, signifying that the individual is free to worship his own god, like a king. In the past, a small stain of this kingly purple was incorporated into each prayer shawl.

❧❧❧

The purple dye produced by snails is a colorless, viscous fluid that oxidizes into an intensely purple pigment. Usually mixed with mucus, in the water it becomes a sinuous column, undulating enticingly with the currents, luring predators away from the snail. A purple-pigment-producing rock snail in the Caribbean lives on forbidding rocky shores just beneath the water's edge, protected by a pounding surf most of the time. In this sharply defined, narrow habitat it is relatively safe from predation from ravenous enemies below and above. But should the seas become calm, the rock snail's ability to use its foot as a suction cup to cling to wave-dashed rocks becomes less significant. An assiduous student can clamber over the sharp rocks to find and remove the snail from its now vulnerable perch.

One young woman, fearlessly leaping from rock to rock, triumphantly brought a specimen to show me. "Do we have this species?" she asked. "No, and it is hard to find. Make sure to keep it squeezed in your fist, lest it escape," I answered. When she returned to the lab, she was surprised to find her palm covered with purple dye and smelling faintly of garlic. She had found *Thais haemastoma floridana*, the Caribbean version of the Mediterranean rock snail. Its purple pigment was discovered by the Amerindians inhabiting the Caribbean basin. Their technique was to rub thread across the foot of the snail, thus staining it purple. The thread was woven into fabric.

Too bad that purple dye is no longer valuable, being replaced by synthesized pigments derived from coal. But these chemicals can irritate the eye. Snail-derived purple dye is not as irritating as the man-made version. It even has an antibiotic effect. Next time you see a debutante dressed to the

PLATE 15

A. CONE SNAIL, *Conus geographus*, attacking a goby with its venomous harpoon-like radula. Cone snail shell to 10 inches long, glossy with brown blotches.

  i. Approach: the goby shows no fear at the slow-motion approach.
  ii. Attack: the proboscis is extended and the harpoon-like radula is plunged into the side of the goby. Venom paralyzes the goby before it can swim away. The radula is often torn free. Some cone snails have a quiver of many "harpoons."
  iii. The cone snail ingests the goby through a greatly enlarged mouth.

B. FLORIDA ROCK SNAIL, *Thais haemastoma floridana*. Gray or yellow with brown spots, to 2 inches high. Caribbean Indians passed a thread along its underside to produce purple cloth. A similar snail, *Thais haemastoma*, is found in the Mediterranean. It also produces purple dye and may have been one of original dye-producing snails.

C. MEDITERRANEAN PURPLE PRODUCING MUREX, *Murex brandaris*. Yellow-gray, to 3 inches high. Probably original source of Tyrrian purple dye.

D. WIDE-MOUTHED ROCK SNAIL, *Purpura patula*. Dull gray, aperture lip purple, to 3 inches long. This Caribbean snail produces a purple dye that is a clear fluid until contact with air, when it turns purple and smells like garlic.

E. A SNAIL RADULA is a file-like structure on the tongue (odontophore). Made of conchiolin, the plastic-like substance invented by mollusks, the radula is used to rasp algae off rocks by herbivorous snails or to rasp flesh from prey by carnivorous snails. The oyster drill *Urosalpinx cinerea* extends its long radula through a distinctive hole that it drills in the shells of clams and oysters to feed on the soft tissues. Many evolutionary modifications have occurred, such as the harpoon-like radula of the cone snail.

PLATE 15

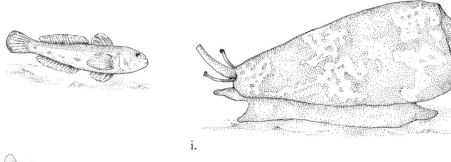

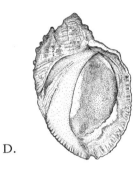

i.

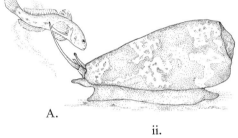

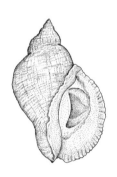

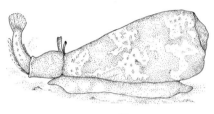

A.

ii.

iii.

B.

D.

C.

E.

hilt, eyelids heavy with purple eye shadow, think of this enticing commercial opportunity.

In the same phylum, Mollusca, the same trick appears to have independently evolved in squids and octopods. When threatened, a sac of black "ink" contracts, forcing out of the anus a threatening black cloud.

⋙⋙ Attack Snails

Why would a snail, encased in a hard shell, evolve a mechanism like purple dye to confuse predators? Sometimes a shell is not enough. In extreme cases another defense system is needed—the writhing purplish mixture of mucus and dye. But this double wammy may not save the snail from that monster of snaildom, the beautiful tulip snail, *Fasciolaria tulipa*, which is one of the most voracious predators of the seashore. It wanders through the shallows, its blood-red body symbolic of its intentions, and attacks everything in its path. Lacking speed, it slithers along on its mucous trail, threatening stationary animals and slower-moving snails. When it attacks, it engulfs its prey with its voluminous foot, immobilizing it. Then it extrudes a file-like tongue, the radula, and files away a hole in the prey's shell. The hole has a characteristic beveled edge.

Some snails, such as the oyster drill, *Urosalpinx cinerea*, are able to produce acidic saliva that dissolves chalky oyster shell, making the filing action all the more effective. When the hole is drilled, the radula is extended through it and into the victim's body, rasping away its flesh. After a few days, the predatory snail moves off, leaving an empty shell.

⋙⋙ Sinister Radula

After the ferocious battle of Guadalcanal in World War II, soldiers were resting before being transported to their next encounter with death. Bored with lying around, a young man wandered the seashore. He found a beautiful, glossy, cone-shaped seashell flecked with brown. He picked it up and was going to place it into his pocket when a barbed needle lanced into his palm. A few minutes later he went into convulsions and died. He had picked up the poisonous cone snail, *Conus geographus*. The snail killed him with a poison dart.

Cone snails are predators on other snails and slow-moving fishes like gobies. The strategy for attack is to move toward the prey animal, following its scent trail. Slowly the snail approaches a stationary goby. Too large to be afraid of the snail ponderously moving toward it, the goby doesn't respond—until it is too late. The cone snail extends its proboscis and shoots out its radula, which has evolved from its rasp-like progenitor into a barbed harpoon attached by a muscular tube to the mouth. The goby makes one startled movement and then shudders as the poison in the dart transfixes it. Just as the Portuguese man o' war produces a virulent, paralyzing toxin, so does the sluggish cone snail. For if the fish were to dash off, it would escape. In its agony, the goby rips out the radula, but it is capable of making only one convulsive movement before death. The paralyzed fish is then engulfed by the cone snail's mouth that has expanded into a huge, gaping maw, and is swallowed whole.

Many species of cone snails have several of these expendable harpoons. Others will produce one after the other. Where the poison is produced is a mystery. The venom bulb seems to consist of muscle layers. It is attached to a duct that stores the yellow, viscous venom. The duct opens into the pharynx, which suggests that the poison coats the radula. The venom usually causes bee sting-like symptoms, although some victims report excruciating pain. The area around the puncture turns bluish and numbness spreads from the site, followed by paralysis and temporary hearing loss. Resistant individuals report extreme weakness for weeks after being stung. Vulnerable victims go into a coma and cardiac arrest.

The soldier survived one of the fiercest battles of World War II, only to succumb to the missile of a humble snail.

16

Size Does Count

THE WAVES crash onto the rocks with a lion's roar. They return to the sea with a soughing whisper. In between the to-and-fro lie big green-black rocks covered with filamentous bright green *Enteromorpha* seaweed and the rubbery, tough brown alga, *Fucus*, resplendent with its necklace of oval bladders. The ebb of the waves stretches the seaweeds out to their fullest lengths. Their root-like holdfasts, seemingly embedded into the rocks, clutch the unyielding surfaces desperately. Respite from this violence occurs briefly for an hour or two when the tide goes out, uncovering the rocks. No longer do the waves smash against the rocky microcosm. But with this respite comes ultimate nullification. Organisms living here cannot function in the air and must sleep until awakened by the return of the crashing, oxygen-drenched water to cover them.

Just visible beneath the flailing seaweeds lie white, conical hillocks about an inch in diameter at their bases. Often they form a white encrustation on the rock, rough to the touch and sharp enough to draw blood when a finger or toe is pressed against them with force. Some of these cones spread out from the colony, becoming more dispersed the farther out they live. Close examination reveals that the cones are made up of plates (scutes) that form a palisade of chalky calcium carbonate. At the pinnacle of the volcano-shaped cones lies a trapdoor (operculum). The tide rises. The operculum opens and jointed, feathery translucent appendages extend to sweep the water like hairy arms. Periodically the rhythmic movement stops and the minute planktonic organisms entrapped in the hairs of the appendages are kicked back into the mouth of the animal. For an animal it is. Each scratchy whitish "volcano" contains an animal lying on its back, tossing food into its mouth like a participant in a Roman bacchanalia.

The clue to their position in the animal world is in the tiny jointed appendages extending from the top of the cone. They identify these attached animals as arthropods (arthro = jointed, pods = legs), kin to the crab and shrimp and other fellow crustaceans. Most crustaceans are gonochoristic (two separate sexes). Male crabs and shrimps court females, employing unique rituals (see chapters 6, 19 and 20). Our volcanic animals are barnacles. Adapting to their sessile (attached) life style, they have incorporated both sexes into the same body. But the dictates of evolution do not permit self-fertilization, else the potential for variability, the stuff of evolution, be reduced. Sexual reproduction is still necessary—a male must fertilize a female; his genome must be mixed with hers.

Destiny has posed this dilemma: how can a "male" barnacle inseminate a "female" when they are mere calcareous bumps on a rock, destined to remain forever attached to the surface with an unusually sticky glue?* Visualize this living volcano feeling the procreative urge. The stimulating substance wafting from a nearby "female" becomes more and more powerful. Furtively, the operculum opens wider and a soft, fleshy projection appears. It sinuously moves toward the source of the sex perfume. But the female is a huge distance away, more than a body length. Not to fear— evolution provides. The male's penis reaches across the void and the two calcareous cones copulate. The evolutionary solution to the barnacle dilemma is to have produced the largest penis (per body length) of any animal, bigger for its relative size than an elephant's.

---

* The glue that attaches the barnacle has evolved over millennia to be so complex and effective that modern researchers have sought to synthesize it. Dentists are particularly interested in this very adhesive substance that sets underwater.

❧❧❧

In spring the intertidal rocks are densely covered with a crust of small white cones. In the fall there are only a few large ones left. Barnacles lead a solitary life sweeping the water with their feather-like setae. If they get their only sustenance by sweeping plankton into their mouths, why are they usually found on rocks or pilings inches or feet above the low-tide line? And why are only a few left from the uncountable thousands present in the spring?

Some barnacle species are inundated with life-sustaining water only a few days of the month, during the highest (spring) tides. When exposure to the harsh, drying atmosphere threatens at low tide, barnacles close their door-like plates at the apex, permitting them to store water for hours or days. They go into a form of hibernation until covered with water at high tide. But if these are aquatic animals, why have they evolved to live above the waterline for most of their lives? The answer lies in their hot dog–like desirability as food. The juveniles have fragile shells, and every marine animal finds them delicious and accessible. Flatworms glide smoothly over the barnacle bed, swallowing the young. Crabs, snails, and fish join the feast. Only those barnacles located on the highest levels of the rock, exposed to the air for most of the day, grow to adulthood.

Added to the stresses of a solitary life, barnacles must produce millions of larvae because they cannot move to defend their offspring. These are typical crustacean larvae, the non-feeding nauplius (named after Poseidon's son, the mythical founder of navigation), endowed with a day's energy in a nourishing yolk. Should the frenzied flapping of its antennae permit it to survive the marauding carnivores of the planktonic jungle, this passive, one-eyed, non-feeding stage goes through a massive physical transformation into a post-larva called a cypris that settles on the bottom for life, attached firmly and permanently by the miraculous barnacle glue.

～～～ The Most Disgusting Barnacle of All

All living things must cope with a changing world. Consequently, there is a constant "search" for genetic variations that become new species. To adapt is to survive. To remain unchanged is to become extinct. Some groups (phyla, classes, etc.) can occupy several biological niches. Should one lifestyle become inappropriate, survival depends on a reservoir of forms utilizing different niches. Most barnacle post-larvae resemble the classic crustacean pattern described above.

But one species, *Sacculina*, produces a unique female arrow-shaped post-larva, the kentrogon. The free-swimming arrow finds a crab and attaches itself to the vulnerable membranous joint of the crab's leg (its "Achilles knee"). Then it injects undifferentiated stem cells into the crab that float in its watery bluish blood to settle in the body cavity. These cells reproduce wildly, becoming a small, undetectable, cancer-like "tumor." The mass expands rapidly into an obscene inner monster, occupying more and more of the confined space within the crab's carapace. Then it begins to engulf the very innards of the crab. The first organ digested is the testis, emasculating the male, and all infected crabs become females. Soon the alien's tendrils penetrate every corner of the crab's body, and there is little inside the carapace except *Sacculina* tissue. Yet, enough organs are ignored to allow the crab to live on.

When it comes time for the parasite to reproduce, a knob-like projection appears on the abdomen of the crab. Buried in the knob are two tiny tubules, vaginas. A wandering free-swimming microscopic male metamorphoses into a tiny testis, enters a vagina, and becomes a perpetual hyperparasite (a parasite of a parasite). Typically two males fuse with the female. Constantly producing sperm, they are in-house gigolos. Eggs and sperm unite and endless hordes of larvae are released periodically. The cycle goes on. Like the monster from the movie *Alien*, the invader uses the body of its host as a vehicle for its own reproduction—only there is no Sigourney Weaver to save the crab from a hideous death.

ᴥᴥᴥ The Most Wonderful Barnacle

On Portugal's craggy shores a gold coast yields treasures, as of old, for warriors brave enough to conquer the sea. Vertical cliffs have been carved from the black headlands overlooking vast expanses once contemplated by Vasco da Gama, father of navigation, whose castle broods over the dark, threatening ocean. Huge waves poise over the murky, dangerous ramparts below the cliffs. Then in slow motion, with monstrous grace, they thrust themselves inexorably against the immovable black rock. Their tremendous force explodes in a white froth. Over eons, the rock at the base of the cliffs has been fragmented into a jumble of scarcely visible crags and honeycombed with caves.

It is in this arena that the *guerreiros do mar* fight their battles.* The gold they seek is the valuable goose-neck barnacle, *Pollicipes pollicipes*, a gourmet delicacy called *percebes* (pear-thay-bays). The thumb-thick, finger-long peduncle (neck) is covered with a gray, parchment-like sheath that is peeled off after the barnacle is served, steaming hot, in a bowl. No butter or sauce is used. Afficionados spurn anything that sullies the white cylinders of lobster-like meat. In season, the wonderful taste comes as much from the muscle of this retractable peduncle as from the "caviar" in its deeply buried ovaries.

Over the last hundred years, overharvesting of these delicious animals has resulted in a progression of the colonies down the rocks toward the more dangerous semi-submerged region. Generations of barnacle gatherers have raped the rocks of their living cover, baring the rocky surface to be colonized by slippery algae. The barnacles are deprived of much of their rocky footholds and survive only in the surf region, forcing the warriors to clamber into regions where they are often entirely inundated by waves. They disappear for a moment then miraculously reappear. Undaunted, using their knight's lances—the steel-tipped sticks called *arrelhadas*—they duel with the barnacles tenaciously attached to crevices in the rocks. Daily,

---

* A wonderful book of photographs depicting Portuguese barnacle gatherers is *Guerreiros Do Mar* by Joao Mariano (Lisbon: Grupo forum, 1998).

these conquistadors return to the haven of the shore laden with their valuable treasure, destined, as in olden days, for sumptuous meals in palaces—gourmet palaces, elegant restaurants where barons of trade feast on them. Occasionally, one of these warriors is defeated by the enemy. His body may be lost forever in the deep.

Another goose-neck barnacle, *Lepas anatifera*, lives not on rocky seashores like its brethren, but only on floating flotsam and jetsam. It has forsaken the stability of the shore to form colonies on a floating bottle or a branch of bamboo. These oceangoing wanderers are carried aimlessly with the currents, living out their lives on their unstable platforms, free of predation but at the mercy of the elements. Other, more conventional, conical calcareous barnacles also sail the seas. They live on specific hosts. One species, often six inches across at its base, is found on whales; another on sea turtles; another on blue crabs. The whale dwellers tend to form colonies with specific configurations that are used to identify individual whales by spotters.

ᔰᔰᔰ Why the Huge Penis?

Why would an aquatic organism maintain the most conservative solution to the problem of insemination, while seemingly more advantageous methods abound in the animal kingdom? Even motile fishes have developed a system whereby they spill their reproductive products into the sea, thus eliminating the need for internal fertilization and the vagaries of finding a mate. Many other crustaceans cast their sperms and eggs into the aquatic environment, producing vast numbers of gametes to insure fertilization. Another option: parasitic flatworms lost inside an animal's body have developed a hermaphroditic reproductive system—both sexes within the same body—but even they cannot self-fertilize and must find a mate. Many species of fishes and mollusks have another approach to the solution: they are protandrous hermaphrodites, first males, then females. Wrasses and parrotfishes are protygynous hermaphrodites, first females then males.

PLATE 16

A. BARNACLES, *Balanus* species, on a rock. Water level is indicated by a horizontal line. Some barnacles are extending jointed appendages covered with hair-like setae from the apex of the shell. Zooplankton, trapped by setae, are thrown back into the mouth of the horizontal body. The barnacle is lying on its back like a Roman at a feast. Feeding can occur only when barnacles are submerged. The tide has come in, inundating some barnacles. A few hours before, they were exposed to the dry air and intense sunlight of the harsh above-water environment.

B. THE MALE ROCK CRAB, *Carcinides maenas*, has been feminized by a *Sacculina* barnacle. The parasite's two ovaries are protruding from underside of abdomen as an undistinguished mass.

C. TISSUES OF THE *Sacculina* have replaced some of the internal organs of the crab. Branching thread-like tendrils expand throughout the crab's body. Only organs that are required for the survival of the crab are not digested. This parasitic barnacle harnesses the crab's metabolic functions for its own use. By allowing the crab to live, the parasite insures its own survival.

D. PENIS OF A MALE BARNACLE extending to a female. It is longer than the barnacle's body. This magnificent demonstration of masculinity is necessary because, like virtually all crustaceans, there are two sexes and fertilization is accomplished by placing sperm into the seminal receptacle of the female. Many barnacles are hermaphrodites, but a male does not inseminate its own female phase, thus preventing self-fertilization. Another male will inseminate the female, providing variability, the stuff of evolution.

Once eggs are fertilized, they hatch into a typical crustacean first-larval stage—a one-eyed, non-feeding nauplius that survives on its yolk until metamorphosing into a carnivorous feeding stage. The yolk-rich nauplius is an essential food for larval fishes and invertebrates.

E. PERCEBES BARNACLE, *Pollicipes pollicipes*, is a delicacy in Spain and Portugal. To 3 inches high. Jointed appendages extend from the top of the largest barnacle, indicating that this is an underwater scene. The black stalk is boiled, peeled, and eaten.

PLATE 16

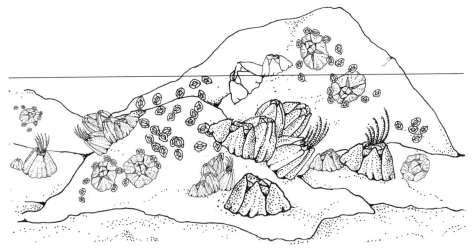

A.

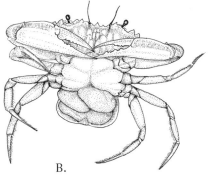

B.

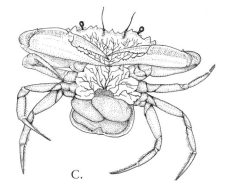

C.

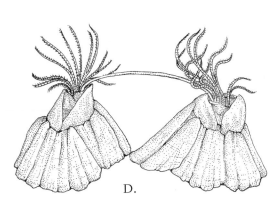

D.

E.

The common slipper snail is called *Crepidula fornicata* because one finds them piled one on another, as if fornicating, four to eight in a reproducing clump. The topmost is a male, the intermediates show gonadal development where more and more testicular tissue is becoming ovarian, and at the bottom is a huge female. The small male, separated from the female by its intermediate "sis/brethren," eventually fertilizes the female by using a long penis. Which species, barnacles or slipper snails, sports the more well-endowed males? This problem has never been resolved. No researcher is willing to spend his days comparing the penis length of barnacles vs. snails. The important issue is that in two diverse phyla, Arthropoda and Mollusca, the huge penis has evolved to solve the problem of internal fertilization of sessile (attached) organisms.* (Thus it is logical that motile snails like sea hares have normally proportioned penises.)

The conservative reproductive style of the barnacle would suggest to a neo-Darwinian that there is some sort of survival value accruing from the retention of a penis in this sessile organism. This seems counterintuitive. The exposed giant penis is vulnerable to predation from the myriad hungry organisms living in the barnacle colony. The potential disaster is a sort of raptorial circumcision. The problem of blindly searching for a vagina seems almost insurmountable.

Perhaps evolution is not directional. Structures and styles may be retained just because an alternative set of mutations has not arisen; a complex structure like a penis would require a vast array of mutations to be replaced. Internal insemination of the female by the male is the norm in many species. Other techniques across the spectrum have appeared, but the default mechanism is transfer of the sperm by an intromittent (penis-like) organ. Some species have a slow rate of mutation. Evolution seems to have passed them by. This ancient solution to survival seems to be so effective that substantial morphological variations have not been selected for.

Maybe the penis *is* the noble organ whose virtues are recounted in song and story. Maybe it is worthy of evolutionary retention.

---

* Technically the slipper snail is motile, but in order to carry out its morphological change from a male to a female, it must remain on the "pile." It is therefore effectively sessile.

17

Fiddler on the Root

It is a balmy tropical evening, music is in the air. The dance begins. He makes his move, a sly beckon. He crooks his finger at her and waves enticingly while rapidly bowing his fiddle. She approaches. His beckoning becomes faster and faster. The music intensifies. He flexes his finger like the villain in a silent movie. If she likes the tune he plays on his fiddle-shaped major claw and is seduced by his gestures, she enters his dark lair. Soon she leaves the nuptial chamber, pregnant. Deep inside her body lies a packet of sperm to be saved for future use.

She resumes her usual behavior, straining mud through her mouthparts, removing and swallowing the rotting organic material contained therein and spitting out little balls of now pristine sediment that lay strewn about her burrow. She is hungry, the metabolic needs of the tiny lives buried within her demanding nutrition. Home to this female is a hole in the mud surrounded by these spitballs.

❧❧❧❧ The Dance

Mating in fiddler crabs, *Uca* species, has evolved into distinctive behavioral patterns that function to keep the gene pool of each species separate. The crabs often are found in marshes or in the soft sandy mud among the roots of mangrove trees. It is frequently impossible to identify what particular mechanism one species uses to draw sustenance from the same environment as other species, but separate their individual identity they must, or else lose the battle for survival as a unique species. Each species "splits the niche" in its own characteristic way.

In a Jamaican mangrove forest there is a sandy patch no wider than a livingroom where five species live. We call it "Crob City" in the Jamaican patois. The crab species can be differentiated from one another by size and distinctive color pattern, but often these characteristics overlap. The definitive characteristic of each species is the tooth pattern on the dorsal finger of the "hand" of the front claw of the male. The claw is called the major chela, and it is swollen and ludicrously huge, three or four times larger than the body. It is not found on the female. She is a two-fisted feeder, using both tiny anterior claws to delicately pick up detritus-laden mud particles for straining with her mouthparts. The male must sacrifice one of his anterior claws to be the "fiddle," devoted exclusively to performing distinctive mating and aggressive behavior displays. The waving-claw mating pattern attracts females of only a particular species, assuring that the genes of that species are perpetuated in their entirety and not mixed. A mixture of two species is called a hybrid, and the new genetic assemblage is not likely to have the survival value of the original species that has evolved over thousands of years to fit into its characteristic niche.*

The mating behavior pattern of *Uca maracoani maracoani* is depicted in the illustration. The male employs a characteristic waving pattern, increasing in tempo as the female approaches. He then plucks at the female's carapace with his small claw. Finally he mounts the female, turning her over and inserting the sperm packet into her gonopore chamber.

---

* If a male three-toed forest unicorn species mates with a female two-toed prairie unicorn species, it is unlikely that the foals will be able to compete with their father's species in the forest or their mother's species in the prairie. The hybrid genome will tend to die out with the maladapted offspring.

~~~~ Crab Combat

The major chela (huge front claw) is also used for a threat display to prevent other males from entering the crab's territory. Many combat patterns are ritualized, for in most territorial animals the energy consumed by battle and the potential for irreparable harm are avoided by a pattern of "huffing and puffing." Fish defend their territories by assuming a pugnacious stance, flashing their flanks, and darting toward their opponent, who often flees. The rule seems to be that the defender usually chases off the interloper, except when much smaller or less energetic than the challenger.

There are fifteen actions of ritualized aggressive behavior that have been recorded in fiddler crabs, most of which involve rubbing the major claw against the opponent's claw. When threat displays are not adequate, conflict results. The aggressor is termed the "actor" and the defender is the "reactor." When the actor approaches the reactor, he attempts to push it backward, often grasping his opponent with the major claw and flipping him. Another pattern involves withdrawal of the reactor into his burrow, extending his major claw part way out of the entrance. Sometimes the reactor pushes a smaller actor back into its own burrow.

Some males wander through the colony, fruitlessly battling successful males. The frustrated "aggressive wanderer" must continue on through the colony to initiate other territorial battles. To lack a territory is to be denied mating, for a major component of the mating ritual is possession of a burrow.

~~~~ Noisy Nuptials

Fiddler crabs can produce an astounding variety of sounds, a colony sounding like a veritable orchestra. Crickets produce their sounds by rapidly rubbing one leg over a file-like, rough-surfaced organ on the other leg. The

PLATE 17

A. Mating behavior of the fiddler crab, *Uca maracoani maracoani.**

  i. As the female approaches, the male begins to wave slowly and horizontally. When she nears, the waving becomes circular and very intense. In other species, male courtship behaviors include curtsying, drumming, and a variety of other sounds.

  ii. The female enters the burrow. Inside the burrow, the male plucks at the female's carapace with his minor chela (small claw), completing the courting sequence.

  iii. Copulation doesn't occur on the surface in this species. Species that mate on the surface perform this ritual: the male climbs the female's carapace from the rear. The female, when receptive, remains in a more or less resting position with her legs bent and held close to the ground. The male's walking legs make stroking or tapping motions. After seconds or minutes of stroking, the male turns the now-quiescent female upside down and the abdominal flaps of both individuals touch. Only the tips of the gonopods (male organs) are inserted into the gonopores. Crabs can remain in the mating position for an hour or more.**

B. Male Burger's fiddler crab, *Uca burgersi.* Huge major chela (large claw) is much larger than body width. Body ⅜ to ½ inch across. Major chela are 1½ to 2 inches long. Brown, tan, or gray; sometimes with reddish hue and brown marbling. The large claw is white or tan, making it conspicuous to the female. Often one can be standing in a fiddler crab colony and see only the males, their conspicuous claws making them obvious. This makes males subject to more predation than females. What effect does the resulting skewed population have on reproduction?

C. Female Burger's fiddler crab. Her claws are petite and equal in size. The balls between the male and female are composed of sand, strained by modified mouthparts to remove organic matter. Except at low tide, a colony appears as a sandy or muddy area riddled with holes surrounded by eighth-inch balls of sand, often near red mangroves.

---

* after Jocelyn Crane, *Fiddler Crabs of the World* (Princeton, NJ: Princeton University Press, 1975), 397.

** Ibid., summarized from p. 504.

PLATE 17

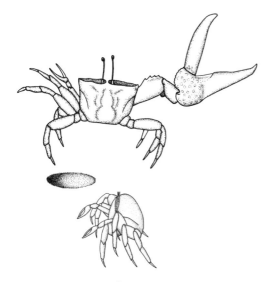

A. i.

A. ii.

A. iii.

B.

C.

rough surface is called a stridulating organ and stridulation is analogous to producing music by moving a bow over the strings of a violin. Fiddler crabs possess no such organ. In general, the small or large claw is rubbed against a variety of carapace structures. Most important are the suborbital crenelations. These are bumps under the eyes that resemble the square, comb-like structures on top of a castle tower. The "leg wag" is performed by rubbing the next-to-last segments of two legs together, producing a low-pitched sound. During courtship at night, the male occasionally produces this low sound to attract a female to his burrow. He then "curtsies" and, at the conclusion of this ritual, the female follows him into his burrow if she is impressed. This same sound is used as a rejection warning by unreceptive females. Interestingly, mating behavior bears a marked similarity to aggressive behavior. (Anyone who has seen cats mate can attest to that fact. Female sharks often bear scars from the bites of males produced during copulation.)

The variety of sounds that have been recorded as emanating from fiddler crabs includes drumming produced by rapidly tapping the tip of a leg against the carapace, and bubbling, when the crab literally blows bubbles inside the gill chamber to produce a rhythmic sound used as a warning. A mysterious sound emanating only from the burrow is called "honking."

❧❧❧ Wayward Females

Some females are wanderers, too. They will approach males in the colony and elicit frenzied waving. But the female will select a male only if she is receptive. If not, she will veer through the colony, causing quite a ruckus, leaving a trail of males waving excitedly from the mouths of their burrows. When she is receptive, she will choose a male and enter into precopulatory display behavior. No one knows what initiates her receptivity. It has been suggested that pheromones may be involved. Another hypothesis is that the female's receptivity increases the closer she is to molting because insemination can occur only when she has shed her protective carapace.

Most data show a relationship between fiddler crab activity and environmental factors, such as length of day, air temperature, phase of moon and tides. But it is the tides that dictate daily behavior of the crabs. They remain in their burrows when covered with water and emerge only when the tide is at its lowest, behaving like terrestrial animals (feeding, mating) for only a couple of hours at each tidal cycle. In fact, so important are tides that a captive fiddler crab can be placed into a covered box and carried miles away from the site with no clue as to the movement of the sea. Its activity cycle will peak in accordance with the time of low tides at the place of origin, though tides change progressively every day.

18

Beware the Duppy

THE FULL MOON creates a silvery sheen over the vegetation. The warm, sultry seaside air hangs heavy over the nearshore scrub forest. A scrabbling sound is heard—first a whisper, then becoming a susurrus of sibilant sounds. Tens, then hundreds of tapping footfalls become louder and louder until the soft sound of the surf is suppressed under the rattling of pointed feet on the dry fallen leaves of the forest floor. Ghostly gray shadow-like forms glide through the vegetation like huge spiders. The "duppies" are on the move. In Jamaica, African ancestry is not far below the surface, and duppies are everywhere. They are dangerous, ghostly, spirit-like memories from West Africa inhabiting dark, scary places.

The dimly seen massed gray ranks are part of the nocturnal nuptial migration of the great land crab, or duppy, *Cardisoma guanhumi*, called the *corrida* in Puerto Rico. But instead of the Spanish bulls of Pamplona, hordes of huge crabs move across the roads of seaside towns toward the ocean. To carry the Spanish metaphor further, during this season many fiestas occur in Puerto Rico, with a land crab stew, *juey* or *cangrejo*, as the pièce de résistance—a kind of West Indian turkey dinner. But every once in a while someone dies from eating the crabs. There is no inherent toxicity in the crab's flesh, but poison is acquired from eating the vestiges of the crab's last meal, a "death apple" (see chapter 4). Knowledgeable crab catchers place these strictly herbivorous crabs in a "corral" and feed them safe foliage for a few days so that they are purged of the remainder of their deadly diet.

June, the month of marriage, seems to have been designated by the gods as the time of procreation, and it helps when the romantic full moon is shining. One wonders whether or not human cues for romance are reflections of biological signals so commonplace among primitive marine animals. Perhaps our primeval ancestors may have had to return to the sea whence they came at crucial moments like June's highest-of-full-moon spring tides. Perhaps we inherit inchoate directives resembling those driving more primitive animals like land crabs to mate. The June moonlight filters down into the cavernous crab burrows, some a foot across. The burrows descend to the water level—as deep as six feet—far from the sea, for survival depends on high humidity that facilitates the passage of carbon dioxide and oxygen through the moist membranous gills.

The crabs emerge from their burrows, almost in unison. The males wave their formidable six-inch claw, but are ignored. A more important imperative drives the females on to the sea. The abdomen of each female is distended with a blackish mass of fertilized eggs resembling caviar. And caviar the eggs will be, to the horde of waiting marine predators.

Land crabs have invaded their stressful marginal habitat as a means of outdistancing their enemies, but have not evolved sufficiently to avoid the primordial need to reproduce in the sea. Such an evolutionary step would require forsaking the crucial planktonic larval development of all marine crustaceans. As the masses of female great land crabs perform their egg-laying ritual, the smaller ghost crabs, *Ocypode quadrata*, stand by their sandy burrows just inches beyond the high tide mark, their ludicrous stalked eyes viewing the scene as if to learn what to do when their turn will come a few days later.

The female lurch toward the sea, their armored bodies burdened with their load of nascent life. They back up toward the water and, for a moment, dip their abdomens into the threatening ocean. In that second, the eggs are released—a cloud of black specks, each a life, each a tempting morsel to the fish and microinvertebrates drawn to the enticing aroma of fresh eggs. Their reproductive ritual over, the females rapidly retreat up the beach to safety. The mass egg-laying process is over by sunrise. The females migrate back to the protection of the dry land, safe from marine predators. Their thick chitinous exoskeleton protects them from all but hungry mongooses, dogs, and man. Fertilization of next year's progeny will occur in the males' burrows.

~~~ The Search for Water

Escape from predators (even larger members of the same species) is the prime directive for virtually all marine animals. The crabs in my aquarium will attack one another rather than eat. Each feeding provokes a mass battle. Wrestling crabs ignore the food and it lies untouched until the fish engulf most of it. Somehow the remainder is sufficient to feed the surviving crabs. Competition between individuals and predation from large fishes and octopods take a heavier toll than starvation on crab populations.

In an evolutionary sense, the land is a safe haven. It is advantageous for some aquatic species to occupy terrestrial niches. Since crabs are recent land invaders, few of their predators have clawed their way out of the sea to pose a threat. Since they successfully crawled onto the land millions of years ago, crabs have explored a variety of terrestrial niches, all requiring yearly egg-laying migrations and all requiring access to water for respiration. Unlike urban humans, who are just now squirming to resolve the water crisis, terrestrial crabs have exhibited considerable evolutionary ingenuity dealing with their need for water. There is a gradation of adaptation in the evolution of crabs onto the land:

- Mole crabs, *Emerita talpoida*, grasp at the land and fail to make the ascent, being relegated to the swash zone of the intertidal. They are buried in the sand as the surf crashes on the shore and breaks in a welter of foam. The crabs position themselves in the sand looking landward. Exposed by the receding water, they extend their feathery antennae to filter detritus from the shifting sand as the waves retreat, defying both aquatic and terrestrial predators.

- Ghost crabs, *Ocypode quadrata*, dig burrows just inches beyond the high-tide mark. Water percolating through the sand satisfies their needs.

- Fiddler crabs, *Uca*, and other mud dwellers face a different problem. Water diffuses slowly through the tiny pores between particles of clay and mud, and the rate of oxygen transfer is inadequate to satisfy their metabolic needs. Oxygenated water comes from twice-daily inundations of tidal water, covering the burrows at each high tide. This limits the habitat of fiddlers to low-lying intertidal mud flats.

- A number of tree-dwelling crabs live part of the time among the fiddlers. Some, like the beautiful orange-flecked spotted mangrove crab, *Goniopsis cruentata*, are struggling to emancipate themselves from their sticky habitat. They spend some of the day in mangrove trees or sandy upland, but must eventually resort to their muddy, moist burrows.

- Nearby, a true tree dweller, the mangrove tree crab, *Aratus pisoni*, has climbed out of the mud, forsaking holes in the ground for the freedom of an arboreal life. It has little need for the safe haven of a burrow, for it is camouflaged to the point of invisibility on the branches of red mangrove trees. Although it inhabits the trees, eating the leaves and rarely crawling onto the land, it will drop into the water periodically to refresh itself and renew the moisture in its gills.

- The half-inch-long marbled marsh crab, *Sesarma ricordi*, solves the moisture problem by living in the leaf litter under the trees. Scrape away some of the thickly layered mangrove leaves, and tiny grayish crabs, each the size of a fingernail, will scuttle away. This species depends on the trapped moisture under the leaves for respiratory water.

- Among the true land crabs, the black land crab, *Gecarcinus lateralis*, and its cousin the mountain crab, *G. ruricola*, have become upland garden pests, establishing colonies of burrows, often overnight, to the consternation of the farmer. These species depend on rainwater for respiration, but are tied to the sea for reproduction.

PLATE 18

A. Male great land crab or duppy, *Cardisoma guanhumi*. Pale gray or bluish. The major claw of the male can be 12 inches long. This male is waving his major claw to entice a female into his burrow. No luck—females (in background) are intent on releasing their burden of fertilized eggs into the sea. No matter how far inland, all crabs require the ocean for the survival of juveniles.

B. Crabs of the seashore.

i. Mole crab, *Emerita talpoida*. Purplish, to 1½ inches long. Found in the backwash of waves beneath the surface of the sand. Facing landward, it filters detritus from the backwash of waves on beaches using its fringed (setose) second antennae that are laid flat on the surface.

ii. Ghost crab, *Ocypode quadrata*, standing next to its burrow on the beach, just above the high water mark. Nocturnal with large black eyes on stalks. Camouflaged with whitish tan carapace, to 1¾ inches wide excluding legs. Runs rapidly, wraithlike, hence its name.

C. True land crabs.

i. Black land crab, *Gecarcinus lateralis*. Body to 1¾ inches wide, excluding the bright orange legs. Distinctive purplish-black marking on its back. Seen here near its burrow on a farm inland. It requires 100 percent humidity and must dig down until it reaches the water table.

ii. Soldier crab or common land hermit crab, *Coenobita clypeatus*. To 3 inches long including snail shell. Large purple claws block the aperture of the shell under stress. No burrow. Can live far inland because it retains water in a chamber, enabling it to keep its gills moist. Rainwater will do, so the crab is essentially free of the need to return to the sea except for spawning.

PLATE 18

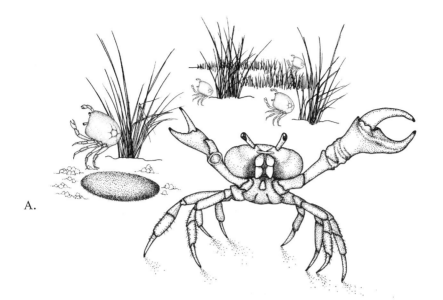

A.

B. i.

B. ii.

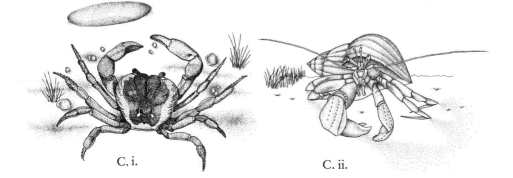

C. i.

C. ii.

❧❧❧❧ Soldiers of the Caribbean

The champion land wanderer is the soldier crab or common land hermit crab, *Coenobita clypeatus*. Using the shell of a dead snail, the crab has a built-in gill chamber in which it stores water. Free of the need for a burrow, soldier crabs can survive for months (years?) on dry land. Consequently, they can be found anywhere on any Caribbean island. So high can it climb that it is found on mountainsides. In fact, I claim the altitude record for soldier crabs. I found a fist-sized specimen near the top of Sage Mountain on Tortola, 1,200 feet above sea level. Soldier crabs move freely on dry land, but they must make the tortuous trek back to the shore when the procreative urge makes its demand. Their dot-like juveniles emerge from the water and hide in seaweed litter close to shore until they can find larger and larger snail shells as they grow to protect them on their sojourn overland. Although they exhibit remarkable habitat adaptability, the presence of juveniles on the seashore gives evidence of their tie to the sea for reproduction.

Soldier crabs stagger along like drunken sailors, sideways, in a comical manner. Because of their clownish mannerisms, they became a prize pet when some entrepreneur discovered their terrestrial tendencies. They were sold in department stores (much like pet rocks). These multifaceted land hermit crabs are the racehorses of the Caribbean. A lean man arrives at a party, carrying a carton of crabs. He draws a circle on the floor (the finish line). Then he releases six land hermits in the center of the circle. Each crab's shell is painted a different color. The gamblers bet on the crab they decide is most likely to succeed. The pari-mutual window closes. The race begins. The "racing man" announces the progress of the crabs in the same tones as a horse race. One or two crabs just sit there, angering their backers, who shout the usual imprecations at them. Old Yeller is nearing the finish line, but Big Red makes a final dash and crosses the line first. Everyone hollers. The winners collect their loot.

We kept a soldier crab on the fireplace hearth. I swear it growled at us when we lifted the cover of its aquarium to feed it lettuce.*

---

* My grandson has a newly acquired soldier crab in a cage (the crab was free, but the cage cost a lot). A free sponge is included to provide moisture when wetted. The clever boy calls the crab Bob—include the sponge and the name becomes "Sponge Bob."

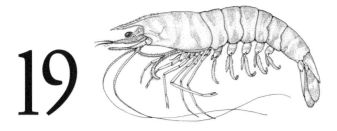

19

The Secret of an Improved Sex Life

I HAVE NO SECRET to improve the *human* sex drive. Perhaps Viagra or eating oysters is the answer. But I can drive a shrimp into a sexual frenzy—at a sacrifice. Just cut off an eye.

How does a one-eyed shrimp mate? Very frequently and with reckless abandon—possibly to the death.

༈༈༈

At the Hawaii Institute of Marine Biology some colleagues were trying to induce mating in the tiger shrimp, *Penaeus monodon*. It is the largest species of commercial shrimp, and if one is willing to ignore its large, dark stripes, it is the most delectable. Thus it is a prime candidate for aquaculture. They put hundreds of these shrimp into a round swimming pool so that the shrimps would "think" they were in the open ocean, swimming around and around the circumference of the pool in a never-ending migration.

Since these shrimp mate only at night, the room was darkened. The quest was to induce mating so that each female would produce hundreds of thousands of gray, fertilized eggs. Normally, the school of shrimps slowly progresses through the ocean, pouring forth uncountable numbers of eggs. These settle to the bottom and hatch into swimming larvae. The plan was to strain the larvae from the water and raise them like chicks on a chicken farm.

But the romantic lighting and make-believe, never-ending ocean didn't work. The shrimps wouldn't mate. Finally someone thought of a drastic technique: remove one eye, stalk and all. A shrimp peers out at the world from beneath protective pockets in its carapace. For maximum peripheral vision, the eye must extend beyond the plastic-like shell. A stalk is the evolutionary solution to this architectural dilemma. Over the eons, the stalk has evolved to house pituitary-like glandular structures. Just as the human pituitary gland produces hormones from two separate regions, so does the shrimp's reproductive glandular system. The human pituitary gland is called the master gland because it controls so many physiological functions. The glands inside the crustacean's eyestalk are multifunctional, too. Growth and sex are intertwined in human and shrimp.

✹✹✹ Ecdysis

Many years ago, I was looking through a tourist guide to that most pleasant, laid-back city, Amsterdam. The dignified discourse called attention to the many night clubs which feature "ecdysiasts." The word troubled me. Subliminally, I knew what it meant, but I couldn't bring the definition into focus. Was it a contortionist dazzling the nightclub audience with astounding rubbery configurations? Was it some public sexual excess? (Since the word vaguely resembles "ecstacy.") The definition came to mind like a bursting bubble. The woman was performing ecdysis—molting. This was a public display of molting—she was shedding her clothes, a stripper!

A female shrimp must perform ecdysis before reproducing. Only then will her naked seminal receptacle, the sperm-receiving structure, accept sperm. Thus, the sexual act requires control of molting. The male cannot cause the female to be ready—no seductive sexual secretion will entice her. She must prepare on her own. He must be available at the moment of truth, when she lies naked, free of her armored carapace, awaiting his advances. The result: the deposition of a sperm packet on her gonopores inside the seminal receptacle.

Thus, shrimps and humans are alike in two ways: molting (shedding one's outer covering) is intimately related to sex, and sexual activity is controlled by hormones.

✸✸✸✸ How Arthropods Strip

If you live in a suit of armor (a rigid exoskeleton), how can you perform bodily functions? (How did knights of yore relieve themselves, anyway?) Arthropods have evolved ingenious solutions to this problem. Some beetles never urinate; they simply store the uric acid crystals under the exoskeleton (carapace) and release them at their next molt. But aside from urinating, the biggest problem of the armored arthropod is growth. Juvenile shrimps grow very rapidly, doubling in size every few days. Each growth spurt takes them to a different size class, one less vulnerable to predators by virtue of its larger size. How can growth be accomplished when the animal is locked into its suit of armor? This process has been refined over millions of years of evolution to be a combination of chemical and physical processes:

1. Growth hormones regulate growth rate.
2. When stored food accumulates in a ratio related to body size, a hormone, ecdysone, signals glands in the epidermis to produce an enzyme, chitinase, that partly dissolves the original armor plating.
3. Beneath the weakened armor the epidermis lays down a new, flexible exoskeleton.
4. An antidiuretic hormone that prevents peeing is produced. Without being able to lose water, the body swells to huge proportions, eventually pressing so hard against the old suit of armor that it splits from top to bottom.

PLATE 19

A. GIANT TIGER SHRIMP, *Penaeus monodon*. Translucent gray with black-striped abdomen. To 12 inches long. Breeds naturally in tropical areas of Australia and Southeast Asia, but now farmed all over the Indo-Pacific region. Adults are usually buried in mud or sand in deep water. Unlike other shrimp, they are primarily predators, feeding at night. Since they live buried in sediment, they cannot be harvested commercially in the wild. Due to their large size and rapid growth rate, they are favored for farming. Two crops of commercial-size tiger shrimps can be raised in a year.

B. CRUSTACEAN GROWTH SEQUENCE.
  i. Dotted line signifies the appearance of sutures at edges of body under influence of hormone from Y gland. Epidermis lays down thin, flexible exoskeleton under original exoskeleton.
  ii. Antidiuretic hormone causes animal to retain water. Body swells, creating pressure that causes exoskeleton to break at suture lines. It is shed. New flexible exoskeleton covers swollen body.
  iii. Shrimp hides until flexible exoskeleton hardens. Thin dorsal line represents flexible exoskeleton covering swollen body.
  iv. Flexible exoskeleton has hardened, signified by thick black dorsal line. Diuretic hormone causes water loss. Body shrinks leaving space for future growth.

C. ANATOMY OF REPRODUCTIVE GLANDS OF CRUSTACEAN. Black bands on eyestalks are X glands. Spots on head near antennae are Y glands. If the eyestalk is removed, the X gland is removed with it, causing a reduction in its hormone. Thus there will be less inhibition of the Y gland and molting and reproduction will occur more frequently.
  i. Eyestalk containing X gland.
  ii. Y gland located on head.

PLATE 19

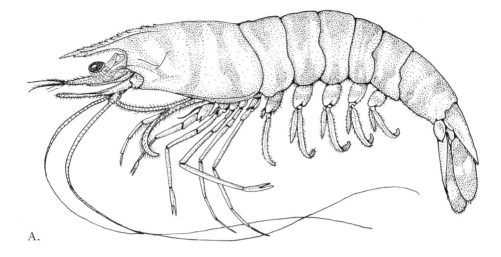

A.

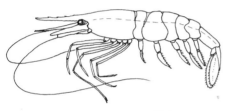

B. i.

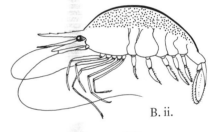

B. ii.

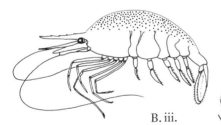

B. iii.

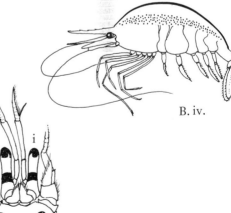

B. iv.

C.

There are lines of weakness around each appendage permitting the extraction of the limbs. Even the lens of each eye is shed.

5. The soft-bodied crab slips out of its previously restricting carapace leaving it perfectly intact. It is wearing its new, flexible armor and hides behind a rock until it hardens. (The molted exoskeleton is such a perfect replica of the original crab that students, intent on catching a specimen, grab what they think is a living crab, only to have it crunch into pieces in their hand.)

6. When the shell is hard a diuretic hormone causes the release of water. The animal pees and pees and the body shrinks substantially.

7. The space between the shrunken body and the new carapace leaves room for growth.

From behind the rock emerges a much larger animal, spiffy with its newly laid-down carapace. (And your average knight couldn't even figure out how to relieve himself.)

≈≈≈ The Dance of the Sex Hormones

If you remove one eyestalk of a crustacean, reproduction will occur frequently. If you are greedy and remove both eyestalks, the female will remain in a sexually receptive mode all the time. She will literally fornicate herself to death.

Crustaceans produce sex hormones in the X gland in the eyestalk and store them in a space misnamed the sinus "gland." Away from the eyestalk, the bases of the antennae contain the Y glands. The interaction between the glands produces sexual activity in an intricate give-and-take similar to the fabled feedback control of the human menstrual cycle.

The Y gland produces the hormone ecdysone that causes molting and also regulates ovary development and food (fat) storage.* Therefore it

* Food deposition and sexual development are related in humans, too. Girls who are undernourished (or athletic, with a high muscle:fat ratio) begin menstruation later than girls with normal amounts of fat and much later than chubby girls, on average.

promotes sexual activity. The hormone from the Y gland inhibits the activity of the X gland.

The X gland produces a hormone that inhibits the production of the Y gland hormone ecdysone, thus preventing molting and, therefore, sex. It becomes dominant as the hormone from the Y gland declines.

If the eyestalk is removed, the X gland will no longer inhibit the production of ecdysone and the female will molt continuously and thus is perpetually available for sex.

20

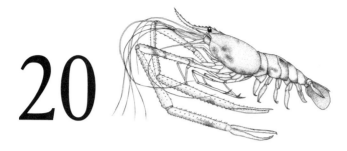

How to Court a Female

*O voluptuous molecule, powerful beyond your
proportions, waft downstream to evoke an
ancient ritual.*

—EUGENE H. KAPLAN

AMONG THE SMELLS of the male prawn's primary food, rotting organic matter (detritus), is the faint suggestion of sexually changed molecules—the aroma emitted when a female prawn is about to shed her carapace. These molecules, present in incredibly small amounts, are perceived by olfactory receptors in the flickering antennules of the male prawn, initiating a genetically dictated sequence of actions. Constantly tasting the life-sustaining water currents, he is drawn to the source of this almost imperceptible molecular aroma.

This stimulus evokes the automatic male response, a search for the source of the smell. Upon finding the female, the male senses that the impending molting process has not yet begun. His immensely long and disproportionately skinny claw-bearing appendages surround the female in an embrace. This is not a symbol of affection, but an instinctive response to the presence of the soon-to-molt female. This male–female relationship is

152

orchestrated by an inherited mechanism to insure the survival of the strongest over time.

Other males, sensing the presence of the female, soon converge on the embracing couple. A larger, stronger male can usurp the position of the defender. The object of this battle for supremacy is to be available at the moment of molting, when the female is receptive, her armor shed. The male turns the female over, inserting a sperm packet into her seminal receptacle. After the sexual union the female ceases to release her seductive pheromone. The aroma is gone; the defeated males slink off. Eventually she produces a new exoskeleton, preventing another insemination.

Afterwards the female produces up to 30,000 eggs, each one passing over the stored sperm packet to be inseminated, then carefully glued to the fan-shaped pleopods descending from her abdomen. The pleopods move to and fro, creating turbulence and oxygenating the eggs. They will hatch in about nineteen days.

ꙮꙮꙮ Farming Giants

In 1959, a man named S. W. Ling, working for the FAO (a United Nations agency to help developing nations find new food sources), decided to investigate the possibility of farming the giant Malaysian prawn, a freshwater lobster-like animal that can attain huge proportions. The record wild giant prawn is preserved in a museum. It is three feet long, not counting the huge clawed legs.

The story is told that Ling had been trying to induce spawning of the prawns for many months, to no avail. Increasingly frustrated, he had taken to working long hours and skipping meals. His wife, growing concerned, would deliver his dinner to him in the lab. One day, Ling, brow furrowed in concentration, tipped over the tray into a large tank containing many adult prawns. Cursing his clumsiness, Ling retrieved the tray, but not before soy sauce spilled, tingeing the water brown. He went home in disgust.

The next morning he noticed that several pairs of prawns had initiated the spawning process. Puzzled, he tried pouring soy sauce into another tank containing many prawns. After a time, they too began spawning. It was clear that the soy sauce contained a substance that induced spawning. What was it: Soy oil? The brown color of the water? Some hidden ingredient? Ling tasted the soy sauce. The predominant taste was saltiness. That was it! He hypothesized that the saltiness in the sauce caused the freshwater prawns to spawn and tested his hypothesis by pouring salty water into a tank containing many pairs of adult prawns with some females about to shed.

The next morning the floor of the tank was littered with huge male clawed appendages, testifying to the carnage resulting from competition between males in that confined space. The females had been inseminated! Ling's hypothesis was validated. Spawning was induced by salty water.

The next puzzle: how did this relate to the prawns that inhabit the freshwater tributaries of the river? He queried local fishermen. They reported that in certain seasons the salty estuary of the river abounded with prawns. The puzzle was solved! In spring the prawns migrated downriver to the brackish estuary to spawn. Much work remained to be done, but the first steps had been taken. Spawning could be induced by adding saltwater to the rearing tanks.

Sensing the huge economic potential of farming these "freshwater lobsters," the states of South Carolina and Hawaii built huge hatcheries to form the foundation of a prawn aquaculture industry. A problem arose: although the females laid eggs that were visible as a gray mass under the abdomen, they invariably lost the eggs after a week or two. Finally someone realized that the eggs had hatched and had become microscopic larvae.

In the wild the larvae join the multitudinous plankton and are swept out to sea, where their numbers are decimated by larger, ravenous zooplankton. Of the hundreds of thousands of larvae released by each female, only a few succeed in making the arduous journey back to their natal streams, only to be caught in swirling currents where death awaits. Even fewer survive farther upstream. Their survival is a matter of luck. Of the millions of prawn larvae leaving the estuary, how many must survive from each mating pair of adults? Only *two*. The prawns solve the problem of replacing each adult pair by utilizing their reproductive energy to create huge numbers of vulnerable offspring. In every case except man, nature keeps populations in balance. If even three juveniles grew to adulthood, this represents a population growth rate of 150 percent. At that rate the population would be more than four times its original number in two generations. After surprisingly few more generations the rivers would be so full of prawns that one could walk across the river on their backs.

In contrast, the related freshwater crayfish produces only a few large eggs that are glued to the pleopods and subsequently cared for, enabling the species to produce the requisite pair of successors by protecting the eggs that hatch as miniature adults. Thus are demonstrated the two strategies that ensure the survival of the species' gene pool: producing huge numbers of vulnerable offspring, few of which survive; or producing a small number of large, heavily yolked eggs that are brooded and provided with protection.

To farm the prawns it would be necessary to solve the problem of defeating nature's intention to produce millions of expendable, feeble larvae. The survival rate of the larvae would have to be increased. To avoid starvation of the millions, as happens in nature, a practical food source for the microscopic larvae was needed. The solution was to feed new hatchlings even smaller, weaker larvae: the just-hatched larvae of brine shrimp.

But despite successes in larviculture, the incipient prawn industry was not to succeed. Hatcheries closed down all over the world. Why? Male prawns grow much more rapidly than females. This rapid growth makes them economically viable. But farm after farm reported unprofitable harvests containing relatively few of the profitable large males. The rest of the catch consisted of uneconomical smaller males and still-smaller females. This impacted the economics so that the farms failed.

Those magical molecules appear again. This time their incredibly minute presence represses male development so that only a few males reach harvestable size. The prawns destined to reach massive size are programmed to be dominant. But most males have the necessary genes to rapidly grow into a dominant male.

Dominant males have huge blue claws. The next largest, but too slow-growing to be economical, are the "strong orange-clawed males." The next are the "weak orange-clawed males." Lowest are the "dwarf males." In the farm pond the dominant blue-claws stake out a territory. Growth of all

PLATE 20

A. GIANT MALAYSIAN PRAWN, *Macrobrachium rosenbergii*. To 3 feet long including claws; mottled greenish gray. Largest males have bright blue major chelipeds (legs with claws). Subordinate males have orange claws.

B. MALE IS "PROTECTING" A FEMALE, waiting for a signal that she is about to molt. Other blue-claw males, sensing the pheromones she is emitting, will challenge the male. The winner will turn the female over and insert a sperm packet into her seminal receptacle. As each egg is released from the female's gonopore, it will be fertilized by sperm emitted from the sperm packet. A dwarf male is waiting in the wings.

C. LARVAL STAGES OF *Macrobrachium* (generalized developmental cycle of all marine crustaceans). All larval forms are microscopic or scarcely visible.

    i. The nauplius is the one-eyed, non-feeding, first larval stage that is full of yolk. It swims with its antennae.

    ii. The zoea, the stage succeeding the nauplius, has two eyes and is able to chase smaller zooplankters.

    iii. Post-larval stage. After many molts, the zoea becomes a juvenile prawn. It goes through successive developmental stages and finally sinks to the bottom to begin metamorphosis into an adult.

D. TWO STRATEGIES FOR SURVIVAL.

    i. Thousands of eggs attached to the pleopods (swimmerets) of a pregnant female *Macrobrachium*. They turn from gray to orange, signifying that the yolk has developed. When they are full of yolk, the fertilized eggs are released into the sea, to be preyed upon by vast armies of zooplankton. Only two need to survive to replace the parents.

    ii. Common southern crayfish, *Procambarus clarki*, uses another strategy. It produces relatively few large eggs (fewer than two hundred or so) and glues them to its pleopods. The larval stages occur inside the egg, and juveniles identical to parents emerge. Also, (iii) juvenile crayfish are protected by crawling onto the mother's back. Survival rates are relatively high.

PLATE 20

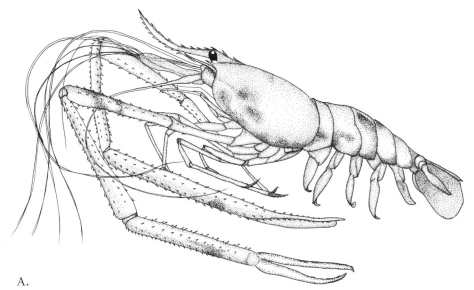

A.

B.

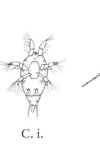

C. i.

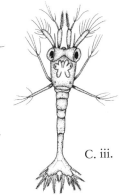

C. ii.

C. iii.

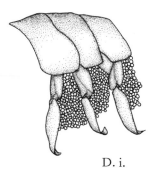

D. i.

D. ii.

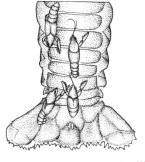

D. iii.

nearby strong orange-clawed males is suppressed by the mere presence of the blue-claws. This hierarchy exists until a blue-claw is harvested or dies. Then, free of the dominance of the big blue-claw, a strong orange-claw immediately initiates the process of becoming a blue-claw.

What is happening? The pheromones are exerting their effect. As long as a dominant male is present, development of the subordinate prawns is suppressed. That explains the small harvest of marketable prawns. Pheromones secreted by the blue-claws suppress the growth of all the other males—and only the few blue-claws come up in the nets. Sadly, the strange caste system of the prawns has brought down the industry.

Incidentally, what causes the continuation of the genetic line of the weakest—the dwarfs? It has been observed that while the huge blue-claw is distracted by the threats from the other large males, a dwarf will sneak under the fierce claws of the dominant male and inseminate the female in an instant of sexual success.

21

The Anti-BLB Club

GRUMPY, FUZZY SCHOLARLY-TYPE was beside himself. He was suspended halfway up a vertical wall made of layers of shale and had reached an impasse. He could go neither up nor down. The ten-foot-high wall was made of inch-thick ledges protruding irregularly from the mesa, and he had run out of ledges. He was stuck. The distinguished professor of physiology and evolution had nowhere to go. I heaved at his bottom and someone pulled upward. He scrabbled onto the flat, dusty, hard surface, prostrate. After a minute he rolled over, tears of exertion in his eyes.

There were about ten people on the top of the mesa, graduate students and faculty—all paleontologists and all within talking distance of one another. Jolly conversations were punctuated with grunts as they pried up layers of the petrified mud called shale. Each layer was about an inch thick, and when pried up, broke into a chunk a yard wide. The technique was to slip the sharp end of a five-foot steel spike into the junction of two layers, and pound at the seam until the layers loosened. The spike was then wedged under the shale and pushed up, breaking off a piece. Then it was turned over to reveal—nothing. But occasionally, even frequently, a professional grunt drew everyone to the digger. He had overturned a piece and there, in all its perfection, was the imprint of a tiny shark, its scales defined, its eyes staring up in a stony stare.

We were at one of those small, scarcely known paleontological sites that abound in the West. This one had an authentic name, Bear Gulch. We were

159

in Montana on a cattle ranch. Our leader was a specialist in the taxonomy of ancient sharks. He had written profusely about shark fossils of the Mississippian period, three hundred million years ago. In those days this was an embayment of what was to be the Pacific Ocean. It was warm and shallow, a perfect pupping ground for sharks. Occasionally a juvenile shark died. It sank to the bottom and was buried in the soft, anoxic bottom mud.

The rich shallow seas were teeming with plankton that lived out their lives in the water column and died, sinking to the bottom to form a layer on top of the mud. Bacteria found them and extracted energy, oxidizing the protoplasm of the dead plankton layer. Reproducing every twenty minutes, the bacterial masses utilized virtually every molecule of available oxygen. Consequently there were few left to oxidize the shark. It remained intact until the mud, under immense pressure over time, became stone—shale. The tiny shark turned into stone along with the layer of mud entombing it. This was the first moment that the sun shone on this animal for three hundred million years!

Profoundly moved, my esteemed colleague and I set to work, grunting frequently—not to prove we were as professional as the graduate students, but because our older bodies would not let us easily lift the heavy layers of rock. Sweating under a glaring sun, we finally succeeded in learning how to use mechanical advantage, wedging up pieces of shale with the best of them. But piece after laborious piece revealed nothing on the underside.

Finally, success! A distinct, curled object, looking like a sinuous peanut, was embedded in the shale. Excited, we picked up the two-foot wide chunk of rock and laboriously dragged it to the "boss." He stared at it intently, then said, "It's a coprolite." Exhausted, we looked at him quizzically. "It's petrified fish feces," he said. We had been laboring to exhaustion only to find ancient fish poop!

We dropped onto the rock floor, laughing a little too hysterically. The rumor spread. We worked until dark, our labors illuminated by a magnificent, luminous sunset. That evening, after dinner, the buttes and mesas rang with laughter as we were presented with our "trophy." (To this day the coprolite lies in state in my class museum.) Bemused by our moment of triumph and thoroughly tired, I crawled into my tent, arranging the nylon floor to carve a flat space through the ubiquitous cow flop. A pat of "prairie pancake" was my pillow.

The next day, one of the professional grunts brought us running. The fossil was of little interest to the group, but I let loose a holler. There on the underside of the rock was a small, perfect replica of a modern-day horse-shoe crab.

❧❧❧ The Crab that Is Not a Crab

The young lady shrieked and ran out of the shallows in a mass of spray. Concerned, we ran to her. She pointed at the water. There, we saw what appeared to be a soldier's helmet. But it was slowly moving. It was a huge horseshoe crab, *Limulus polyphemus*.* Attached to its posterior was an-other, smaller horseshoe. I explained that they were looking for a shallow spot at high tide in which to lay their eighth-inch, grayish eggs. She squirts out the eggs and he, the smaller one, instantly inseminates them. Once they are fertilized, she covers them with a shallow layer of sand and crawls away in tandem with the male. Fish and birds move in for a meal of horseshoe crab caviar. In fact, red knots (birds), in their annual migration, alight after flying thousands of miles, onto beaches to gorge themselves on the eggs, to revive and refuel for the next leg of their migration north. If they do not find a ready supply of eggs, many will not be able to reach their nesting grounds. Thus, a fluctuation in the populations of horseshoe crabs leads to a commensurate variation in red knot populations.

I picked up the two-foot-long crab by its sharp, spike-like telson (tail). It dangled from my hand, twelve clawed appendages churning and book gills flapping, a truly horrendous sight. There is a flatworm, *Bdelloura*, which lives *only* on the paper-thin book gills of horseshoe crabs. I showed the class this white, flat, half-inch worm on one of the flapping plastic-like gills.

Searching for food, the crab bulldozes the surface of the sediment, push-ing up the top layer of sand with the front of its carapace, making a shallow

---

* Polyphemus was the mythical one-eyed giant blinded by Odysseus. The reference may allude to the presence of a tiny dot-like light-sensing structure in the front center of the ani-mal's carapace, suggesting the single eye of the giant.

PLATE 21

A. MATING HORSESHOE CRABS, *Limulus polyphemus*. Dark brown, to 28 inches including tail. Two major eyes are augmented by a concentration of three eyespots (ocelli) that appear as one in the middle front. These ocelli are used for sensing shadows of predators and day and night.

Thousands of pairs of mating horseshoe crabs litter the intertidal zone on remote sandy beaches in the spring. The smaller male is clasping the tail of the female, who is laying up to three hundred eggs in the sand. The male detects pheromones released during the egg-laying process, inducing him to simultaneously release sperm. The tail (telson) is raised when the animal is stepped on, and used to right the animal when it is overturned, but has no aggressive role.

B. UNDERSIDE OF HORSESHOE CRAB. The front two pairs of "legs" are mouthparts: the slender, clawed chelicerae grasp prey and the pedipalps are used for walking. There are only eight true legs. Eight legs, chelicerae, and pedipalps are characteristics of spiders, ticks, and other arachnids, suggesting that horseshoe crabs are closer relatives to spiders than to true crabs.

Because the mouthparts are not available for chewing, spiny "armpits" (gnathobases) grind up soft prey (worms and juvenile mollusks).

A series of thin flaps behind the legs are book gills. They move back and forth, enhancing oxygen transfer into the blood. A flatworm, *Bdelloura*, lives only on the book gills of horseshoe crabs, feeding on detritus that might interfere with gas exchange.

C. CLOSE UP OF GNATHOBASES chewing up a worm. The tail of the worm can be seen projecting from the rear of the gnathobases. The worm will be passed forward, running the gauntlet of successive pairs of gnathobases. The pulpy mass eventually ends up in the mouth (black spot).

D. COPROLITE (ANCIENT FECES), about 1½ inches long, on the underside of a slab of shale. This is an impression; the organic material has long since rotted away.

PLATE 21

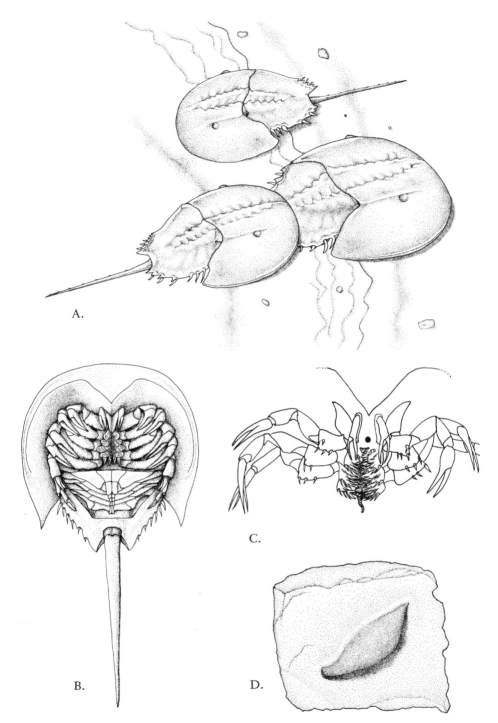

A.

B.

C.

D.

trench. A worm is uncovered. It grasps the worm with clawed mouthparts called chelicerae. But, like all arthropods, the front part of the mouth and pharynx are a tube made of plastic-like chitin, so it cannot swallow large pieces of food. An ingenious solution to the problem: The worm is passed back to the rear legs and is carried forward by creatively named spiny armpits called gnathobases. It is "gnathed" up by these spiny proximal segments of the legs, and by the time it reaches the mouth, it is reduced to chopped meat.

What makes horseshoe crabs different from "real" (crustacean) crabs? The presence of a pair of prehensile mouthparts that project from the front of the body, called chelicerae. Horseshoe crabs use them to grab prey; spiders, to inject poison. Ticks have evolved chelicerae that are razor-sharp stilettos to cut a hole in the skin for the insertion of a blood-sucking tube. These structures cause spiders, ticks, and horseshoe crabs to be called "chelicerates," in contrast to true crabs, which are "mandibulates." Mandibulate mouthparts consist of an anvil-like grinding structure surrounded by three pairs of "chewers" called maxillipeds. In other words, horseshoe crabs are more similar to spiders and ticks than true crabs.

Bad Little Boys

On the way back to the cars, we saw a crowd of kids converging excitedly around a stranded horseshoe crab. Every once in a while one would throw a rock at it, causing an eruption of colorless blood to spurt from the hole in its carapace. It was then and there that I swore in the first members of the Anti-BLB Club. BLB stands for Bad Little Boy, and the members of the club must swear that if they come upon a group of Bad Little Boys throwing rocks at a horseshoe crab, *they will throw rocks at the BLBs*. Boys randomly kill what they fear, endangering the survival of a harmless animal that has existed for more than three hundred million years.

An evolutionary miracle, horseshoe crabs invented an appearance and lifestyle that defied change. Unaltered through time and environmental vicissitudes, its ancestral survival techniques were so perfectly honed that it did not go through noticeable physical changes. When we take offense at what the children are doing, they ask, "What good is that ugly thing?" We try to explain that the horseshoe crab fits into the ecological web in a precise niche. It is hard to explain about the interrelationship between the horseshoe crab and the red knot, a bird. The theory is not absorbed, and when we leave, the kids go back to molesting the animal. Throwing rocks at the "Bad Little Boys" makes a better impression.

In fact, the horseshoe crab is medically important. It would be ironic if one of the little boys would have need of a diagnosis that can only be accomplished through the use of the clear, bluish blood of the horseshoe crab. The blood of this elegant animal, this "blueblood," contains hemocyanin, a bluish respiratory pigment so ancient that it is less effective at carrying oxygen than our red pigment, hemoglobin. The immune system is primitive also. In fact, its response to invading organisms (e.g., germs) is generalized. That is, it produces antibodies to any and every bacterium that might be harmful. Its immune system is not specific to particular kinds of attacker, unlike ours. That makes the blood extremely valuable in modern medicine as a broad-spectrum test for the presence of pathogens. The liquid part of the blood is separated from the antibody-producing cells called amebocytes. These are made into a substance called LAL (limulus amebocyte lysate). Adding this to an unwell person's blood sample produces a visible clot if any harmful bacteria or their endotoxins are present. (Some allergens can also be detected using LAL.)

Horseshoe crabs make good bait for eels and crabs. Baymen in Chesapeake Bay and elsewhere capture horseshoe crabs, pound them into mushy masses, and put them into eel traps, killing thousands. Despite human depredations, despite their vulnerability when mating in the shallows, horseshoe crabs have survived. Three hundred million years of using the same model may allow the species go on through time, but man has rendered extinct many successful ancient animals. The issue is still in doubt. You are invited to join the Anti-BLB Club.

22

Sea Pussy

THE BOTTOM, ONLY six feet below, was barren sterile-appearing sand as far as the eye could see, the only relief occasional tufts of sea grass and chunks of dead coral. No movement. No life. I didn't understand. This was the reputed sea urchin paradise in the echinoderm epicenter of the Caribbean, Jamaica. I returned to the lab and recruited an urchin expert. At his direction the boat revisited that very spot. He dove to the bottom and waved his hand. Presto! A brown, fuzzy urchin appeared. What was the magic formula? What was the secret of finding urchins in this underwater desert? He did his magic again. One wave of his hand and another urchin appeared. I could see bubbles ascending from his mouth. I imagined each contained the word "abracadabra."

Then I noticed that there were barely distinguishable mounds in the sand, boundless numbers of them. I descended and waved my hand over a mound, scattering a cloud of sand. When it settled, there was revealed a brown, oval prickly animal. It was *Meoma ventricosa*, locally called the sea pussy because its fuzzy appearance resembles a woman's pubic region. (I have searched for a more innocuous reason for the name, but have come up empty handed. When looking at the urchin, the name is all too apt.)

Joyously, I employed my newfound skill. Before I finished, I had uncovered what seemed to be about half an acre of four-inch-wide, inch-thick, prickly brown sea pussies, each frantically burrowing, seeking to cover itself with a sandy cloak of invisibility.

Suddenly I heard the characteristic croak a snorkeler produces, a water-distorted call that usually means "come here, I have found something." My guide was pointing excitedly at the bottom. I saw nothing. He was emitting huge bubbles, each a solitary word, "look." I strained to see through my mask's thick prescription lenses. All I saw was a neat row of "toothpicks," whose tips barely projected from the bottom. I marveled at his ability to concentrate, to differentiate these tiny projections from the welter of shapes on the bottom. He waved his hand back and forth. Revealed, in all its glory, was the most magnificent of sea urchins, *Plagiobrissus grandis*. And grand it was. Its foot-long *test* (exoskeleton) was covered with sand-colored stiff spines, like the hair of a blond teenager with a fashionable spiky haircut. From one end, projecting upward, were about twenty toothpick-like calcareous spines. It was oval, with a bulge on its bottom surface, clearly identifiable as a mouth, and another, near the posterior, the anus. Amazed, I gasped, sputtering as water entered my snorkel. I was looking at the rarest, most legendary of Caribbean sea urchins—also the most highly evolved.

ꕔꕔꕔ Urchin Evolution

All echinoderms are shaped like a pie and dissected by five (always five) grooves called ambulacra, from which myriads of tiny tube feet (podia) project. The animal walks along on these feet. When it buries itself, each minuscule tube foot shovels the sand away and the urchin settles into the sand and slowly becomes invisible. But being shaped like a circle has its disadvantages. Glacially slow burrowing exposes an urchin for too long to its enemies. It is advantageous for the urchin to evolve a pointed or wedge-shaped "front" so as to plunge through the sand and more rapidly escape. The evolution of shape in sea urchins is one of the most clearly spelled-out demonstrations of animal adaptation. The sequence of shapes is simplified by the fact that it is not necessary to go into the fossil record. *All of the shapes are demonstrated by living animals.* With a little imagination one can visualize the evolutionary steps.

The rock-dwelling urchins and those exposed on sandy surfaces are globose, projecting outward a symmetrical armament of sharp spines capable of warding off attackers in any direction.

What may be the next step in the quest for safety is to develop a sharp-edged, wafer-thin body and fuzz-like bristles, still retaining the circular body, such as is found in sand dollars. The reduction of spines into a sandpaper-like surface and its sharp shape enables the animal to slip beneath the sand quickly, leaving no characteristic mound. Caribbean sand dollars, *Mellita quinquesperforata*, have evolved another safety mechanism. Five oval holes in the body, called lunules, act as spoilers. When a huge wave uncovers the urchin and carries it shoreward to certain death, the lunules create turbulence, and instead of skimming along with the wave, the sand dollar sinks like a stone.

The next evolutionary development is to evolve a front. In sea urchins, this is a stumbling, recent evolutionary strategy, one that has long conferred superiority to our kind. For to have a front means it is possible to accumulate sensory structures: eyes, nose, mouth, and, thereby, to be able to chase prey. Of course, the bumbling sea urchin has no need to chase prey because it eats rotten fragments of animals and plants, called detritus, mixed in with the sand. But it needs to avoid predators. Thus to subterranean urchins, a front has survival value.

The thick, circular sea biscuit, *Clypeaster rosaceus*, lays partially buried in the sand, camouflaged by a coat of seagrass leaves. It has a central mouth, but the anus has migrated to one end, suggesting a posterior.

The sea pussy has an oval shape. A skeletal projection on its underside faces forward, suggesting a gaping mouth, and the anal opening is a hole in the rear margin.

Plagiobrissus has solved the evolutionary problem of front and back. On its underside projects a cavernous opening, the "mouth." This makes this region a front. Towards the rear is another opening, projecting backward, the anus. Between the two is a brushlike mass of spines upon which it walks. Surrounding the mouth and anus are many thick, blood-red tube feet. Is this urchin so advanced as to have evolved the red respiratory pigment, hemoglobin? Have the thick podia evolved the function of sweeping food into the mouth and feces from the anus? Clearly this "super sea urchin" may have reached primacy among urchins.

Its rarity has deprived this remarkable animal of a common name. I have assumed the responsibility of naming it The Great Red-Footed Urchin* in deference to its magnificence.

❧❧❧ The Crown of the Antilles

The most fearsome aspect of the tropical underwater vista is not the occasional shark, nor the silvery barracuda that circles the diver, glaring at him with baleful eye. No, the most frightening monster of the deep is the long-spined black sea urchin, *Diadema antillarum* (Crown of the Antilles). Six-inch-long spines, covered with thin, transparent poisonous skin, are able to rotate, so that when the diver approaches, he is faced with a formidable bundle of razor-sharp needles. Sometimes the bottom is so covered with these scary black urchins that it seems inevitable that the snorkeler will be impaled, but often in the glare of day they have hidden themselves in a niche in the reef, safe until dark shadows signal that it is time for a night's foraging. At dusk, holes in the reef slowly give up their black shadowy inhabitants, until a belt of urchins flows along the length of the reef. During the day the depredations of the long-spined black sea urchin are visible as a wide, bare, desert-like band edging the reef. The herbivorous urchins have migrated each night about twenty feet from the haven of the reef to feed on sea grass and algae. They dare not move any further, for predators are waiting. How can hunger drive fish or crab to penetrate the impregnable fortress of poisonous spines?

The queen triggerfish, *Balistes vetula*, outsmarts the urchin in a demonstration of the superior flexibility of vertebrate over invertebrate. First, it creates a wave by flexing its tail sharply. This upends the urchin. Then the fish viciously pounces on the urchin's exposed bottom that is protected only by short, blunt spines. *Diadema* must walk along the bottom on truncated tube feet, and the bottom spines must be equally stubby. Very long spines would cause the animal to teeter along, tumbling end over end with

---

* Eugene H. Kaplan, *Field Guide to Southeastern and Caribbean Seashores* (Boston: Houghton Mifflin, 1988), 356.

each bottom wave. The triggerfish munches on the entrails of the no-longer threatening urchin, leaving, as testament to its superiority, an upside-down test surrounded by a halo of long spines. Even *Diadema*—this sinister, seemingly impregnable sea monster—has enemies.

When dawn's diaphanous light filters to the bottom, some long-spined black sea urchins fail to heed the warning. Caught far from their dark crevices, they creep toward the reef. In the sun's glare every shadow suggests a hole in which to hide. The urchin's surface is covered with sensory ocelli (eye spots), each capable of differentiating light from dark. Thus it is able to tell dawn from night (and to aim its spines at a diver's threatening shadow). The primitive ocelli are unable to perceive images. But the urchin can detect a dark shadow resembling a hole and moves toward it, unable to discern that it is another black urchin. When the urchin reaches its ostensible haven, myriads of spines greet it, and the two urchins spend the day jousting for protection. Other urchins move toward the double-size black blur. The mass becomes larger. Finally, all the nearby urchins (usually about twenty) form a characteristic wedge-shaped quivering group. It seems that the most efficient arrangement for the urchins is a triangle.

❧❧❧ When Tragedy Strikes

Inevitably, a student is impaled by a hidden urchin. The poisonous skin covering each spine produces pain similar to that of a bee sting. The students' reactions vary from stoicism to crying.

There are several methods of dealing with the broken-off tips of the spines, which appear as a group of black dots under the skin. The method varies with the student. If he or she has been obnoxious, the "beer bottle" method is used. Each method is described below. All begin with bathing the injured region with meat tenderizer. This contains papain, a protein-digesting enzyme from papaya fruit. Thus the proteinaceous poison is rendered harmless.

Method 1. Shut the lights. Light candles. Mutter incantations in as convincing a manner as possible. Tilt the candles and pour melted wax on the black dots under the skin punctured by spines. Let the wax harden. Rip it off, reaching a crescendo in your chants. More often than not the wax will penetrate the holes produced by the spines. Turn over the wax wad and the black dots will be visible, embedded in it. Place antibiotic salve on the wounds.

Method 2. Send the student into the bushes to pee on the spines. How he or she accomplishes this remains unseen. The theory is that the acidic urine (pH 4.5–6) neutralizes the alkaline toxin (?) and dissolves the chalky calcium carbonate spines. Wash and apply antibiotic.

Method 3. I have used this only once. On Bonaire, noonday temperatures reach 120°F. Only mad dogs and college students venture out in that noonday sun. Consequently it is necessary to forbid students to walk to town around noon. Two rotten kids not only went to town but dove off a cliff into a mass of *Diadema*. They returned with hands speckled with broken-off spines.

This called for drastic measures. A time-honored method of dealing with subcutaneous spines is to break them up into powderlike fragments that soon dissolve. Gathering the class around, I took a beer-filled bottle and rolled it over the afflicted palms. Everyone winced. The macho kids forced themselves to be stoic. The method worked. By evening the spines were no longer visible.

These methods have the same efficacy as any other folk medicine; the suggestion of a cure removes the pain. It also creates memories that will provide bragging rights for years.

﹋﹋﹋ Haven or Heaven

By virtue of their formidable spines, long-spined black sea urchins provide a moveable fortress for a variety of vulnerable species. The urchin crab, *Percnon gibbsi*, has an ephemeral relationship with the urchin. It is often found under a *Diadema*, gold-striped legs glowing from under its formidable protector. Wherever the urchin wanders, the crab is sure to go, making

PLATE 22

A. Underwater scene showing mounds of sand on the bottom. Bristles of sea pussy urchin, *Meoma ventricosa*, are barely visible under the sand. Vast areas of lagoon bottom consist of bare sand with turtle grass and macroalgae in small clumps, only to be relieved by *Meoma* mounds.

B. Upper surface of sea pussy, reddish-brown, covered with short bristles. To 5 inches in diameter with four angled grooves and a central narrow depression.

C. Underside of sea pussy showing tracts of bristles. Anterior mouth projects downward as an unmoving, scoop-like lower jaw. The body is elongated and the anus is posterior, suggesting an evolutionary step toward bilateral symmetry.

D. Long-spined black sea urchin, *Diadema antillarum*. To 6 inches in diameter including black calcareous spines covered with poisonous skin. Ocelli (eye spots) at the base of spines distinguish daylight from shadows and darkness. Spines can rotate to point at the shadow of a diver. *Diadema* migrates from coral crevices in the dark of night.

E. Razorfish, *Aeoliscus strigatus*, white with longitudinal black line. To 5 inches. Although free living, it finds safety camouflaged among the spines in times of danger. The other fish depicted is a clingfish, *Diademichthes lineatus*, a true parasite. It lives on *Diadema*, feeding as a juvenile on the surface structures of the urchin and as an adult on its tube feet.

Many species, from a variety of phyla, find refuge among sea urchin spines. A tiny crab, *Dissodactylus*, feeds on detritus on the surface of the urchin. The urchin crab, *Percnon gibbesi*, often hides under *Diadema*, but the relationship is ephemeral and the crab survives well under rocks in the spray zone of rocky shores.

PLATE 22

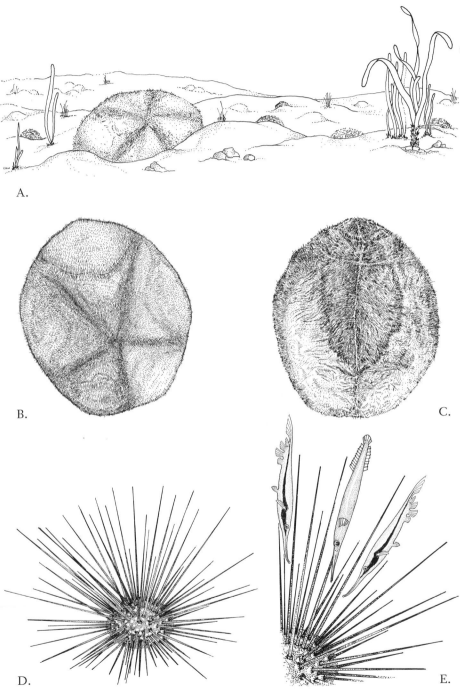

A.

B.

C.

D.

E.

feeding forays when predators are not near. But one can often find the crabs wandering the bottom on their own. They do not seem to have a well-developed relationship with the urchin.

Not surprisingly, several species of remarkably adapted fishes live among the *Diadema* spines. Long and narrow, they float vertically, almost invisible in their permanent refuge, peering out from their black toxic fortress, waiting for the occasional planktonic passerby. They dart out for the attack—then rush back. The razorfish, *Aeoliscus*, camouflages its body with a black stripe on a whitish, nearly colorless body. The black vertical stripe perfectly mimics a black spine.

A more intimate and permanent relationship converts "haven" into "heaven." Species of pea crabs, *Dissodactylus*, tiny eighth-inch white, almost invisible crabs live among the spines of several species of urchin, obtaining protection and sustenance. They must live on their particular host urchins, feeding on detritus that accumulates on the surface of the urchin, and hide among its spines. The question: is the removal of the detritus beneficial to the urchin? In that case the crab/urchin interaction would be an example of mutualism. If the crab occasionally takes a nip of the urchin's skin, the crab is harmful and is labeled a parasite. If, however, the crab does not help or harm the urchin, and it clearly benefits from the relationship, then it is a commensal. No one knows the answer to this question concerning these crabs. *Dissodactylus crinitichilis* lives only on the six-holed sand dollar. *D. mellitae* lives in the slots and underside of the five-holed sand dollar, *Mellita*. Are they helpful, harmful, or neutral?

Undoubtedly a species of pea crab lives among the spines of the long-spined black sea urchin. No one, to my knowledge, has had the courage to search for it.

The Monster Returns: The Triumph of Good over Evil

In the 1980s long-spined black sea urchins began to wither away and die in massive numbers. First their populations drastically declined near Panama.

Then death spread ever-outward from that locus. Nothing else died, just this species of urchin. It became apparent that a plague was afflicting them. Hotel owners noticed this and smiled. Scuba-diving operators were happy. Gone were the dreaded monsters that impaled their guests on many occasions.

Then everyone noticed that the reefs were being smothered by crustose algae. The crusts suffocated the corals by preventing life-giving light from reaching the polyps. Then sponges and leafy green and brown algae covered everything, preventing coral larvae from settling on the bare, hard surfaces they need. The reefs became algal reefs. Divers and hotel guests declined in numbers. Who wants to travel thousands of miles to see algae?

What could be done? No one knew the answer. Then someone tied together the decline of urchins and the algal plague. The urchins kept the algae at bay by grazing on the reef. Before an algal turf could develop, before the coral could be covered with algal crusts, the ravenous black-spined herbivores chewed the algae into harmless populations. The urchin plague was on such a massive scale that any human effort would be inconsequential. The complex ecological web of the coral reef had been disrupted. It was evident that the long-spined black sea urchin was necessary for the survival of the reef. What was feared and hated now was loved. Now the urchins have returned. They are much appreciated.

> *Don't it always seem to go that you don't know*
> *what you got 'til its gone?*
> —JONI MITCHELL

23

Debunking the Big Lie

THE RAKISH BOW of the old wooden sloop cut through the water, creating a small, glistening bow wave. The gaff-rigged mainsail ballooned outward in response to the freshening breeze. The boat listed slightly to starboard in response to the combined pull of the sail and the "net" being dragged along the bottom.

At the end of the run, the captain signaled the helmsman to turn "hard a' lee" and the boat swung away from the wind, sails luffing impotently. The "net" lines sagged as the boat's forward velocity waned. The captain signaled a crewman to operate the winch. Boom groaning, the load broke the surface with a gigantic splash and was swung aboard, dripping huge rivulets of water, soaking the coarse wooden planks of the deck. The boom jerked up and down, spilling the writhing catch into a mound of purplish, spiny arms. The "net" was a huge white mop, with six-foot-long, half-inch-thick ropy strands.

The catch, spines entwined in the mop, was a purple mass of sea stars (starfish), *Asterias forbesi*. Each shake of the boom freed some of the animals, which soon were strewn haphazardly over the deck. A crewman swept the scattered, rigidly contorted animals into a neat pile. The boat slowly drifted leeward on the rippling Oyster Bay waters. The captain and crew gathered around the mound in a deadly ritual. Long, sharp knives in hand, they slashed at the sea stars, a gleam of revenge in each man's eyes.

The predators endangering their living as oystermen were being destroyed. They worked systematically. Halved sea stars fell to the deck, spilling vital fluids, rendering men and deck slippery and slimy. Finally, the mound was reduced to pieces—individual arms, halves, chunks—and the pile was swept over the side. The captain signaled for another run and the operation was repeated again and again throughout the day.

ᵕᵕᵕᵕ

Bay bottoms along the East Coast are divided into leaseholds, a legacy from colonial times. Gradually, as oyster populations waned, individual lease-holds were abandoned. The surviving oystermen pored over eighteenth-century maps to delineate the margins of adjacent leaseholds and melded them together into huge underwater estates, upon which they had sole rights to harvest oysters. Areas open to private citizens and baymen became scarce. By the end of the nineteenth century oyster populations on public bottoms had begun to decline. Sailing vessels were replaced by small steamboats, their increased efficiency further reducing oyster numbers. But a few wooden sloops remained. When in full sail, these remnants brought back beautiful images of past full-rigged working sailing vessels.

Unknown to the captain of the aforementioned sloop, his satisfaction with his grisly deed was unwarranted. He had perpetrated a disaster! As the boat drifted leeward, a wide swath of bottom was strewn with halved sea stars. Sensing the body fluids escaping from the injured animals, predatory fish gathered. Toothed mouths ingested the sea stars—and spit them out, spiny surfaces repelling even the most aggressive attacker. Soon the bottom, covered with living halved sea stars, reverted to rocks, oysters, and sand. The sea stars had miraculously disappeared! A close view would have revealed the tips of buried arms projecting from the sand or from interstices in the oyster reef.

After a time, a man-made horde of peculiar-looking sea stars replaced the original halved harvest. These newly refurbished animals could be distinguished from their undamaged brethren by their asymmetry. Two or three normal arms extended from the central disk. But several complementary (to make five), tiny, pimple-like miniatures of former arms rimmed the rest of the disk. The sea stars had performed the miraculous and very effective characteristic mechanism of their phylum, regeneration, and the "pimples" would become full-sized arms.

Unwittingly, the captain and his crew had loosed upon the oyster reef an army of sea stars that was double the size of the original. Their assiduous, daylong efforts magnified the ranks of the sea stars until the oysters were decimated to below commercial levels, delivering a death blow to the industry in many mid-Atlantic bays.

܀܀܀ Unmasking the Lie

Most of the predators on the oyster bed eat the newly settled juveniles. Oysters are vulnerable to crabs, fishes, and other small predators until they reach dime-size, at which time the shell has thickened and hardened. But sea stars are able to feed on the larger oysters, using a technology that humans have discovered and perfected only in the twentieth century. This is a battle between hydraulics and muscle power.

The sea star wraps its five arms around the oyster. From each arm project hundreds of half-inch-long tube feet (podia). Each one has a suction cup at one end and a nose dropper-like bulb on the other. Tiny muscles contract. The bulb is squeezed and water fills the tube foot, extending it until the suction cup lies firmly on the oyster's shell. Then a valve releases the water out of each stretched podium and it contracts like a released rubber band. This creates a tiny pull on the shell. Magnify this barely perceptible pull by a thousandfold, and there is created an inexorable pressure on the shell to open. Once applied, no energy is required to contract the podia. But the huge adductor muscle of the oyster resists the pull, keeping the shell tightly closed. The battle between "modern technology"—hydraulics—versus the ancient defense mechanism—muscle contraction—is joined. Gradually the pressure exerted by the combined pull of thousands of tube feet exhausts the oyster's muscle. The lactic acid produced as part of the energy-providing chemical process builds up and poisons the muscle (just as your biceps will become excruciatingly painful if you hold your arms out in front of you for a long time. They must involuntarily drop after a while. (This was used as a torture technique by the Nazis in World War II.)

The oyster's single muscle eventually weakens, exposing its soft parts. It is exhausted. The stomach of the triumphant sea star is extruded from the mouth in a diaphanous hollow sheet. It forces its way between the separated shells and secretes flesh-digesting enzymes. Over time the oyster is reduced to two empty shells and the voracious sea star moves on.

All of this is a misrepresentation perpetrated on generations of students.

The truth: Long ago a Japanese researcher wired shut some live oysters. He placed them into an experimental tank, then added sea stars. In a few weeks he unwrapped the tightly wired oysters. They were empty! What is the only explanation? The sea stars, unable to open the wired shells, were somehow able to feed on the oysters anyway. It turns out that it is unnecessary to pull the shells apart. The two oyster shells, top and bottom, do not close perfectly. Invariably there is a fissure, a flaw in the closure. The sea star is able to insert its filmy stomach adequately enough to enter through the minuscule gap. Once inside, it secretes its enzymes, gradually digesting the oyster's adductor muscle. The shells gape, leaving the oyster's body vulnerable.

ꙮꙮ A Sad Story

Long ago a friend of mine was researching an important problem: oysters release clouds of sperm and eggs that fuse to produce a typically molluscan larval form, a near-microscopic sea-going vacuum cleaner called a veliger. It eats the tiniest algae by sucking them into its mouth by means of two perpetually beating circular masses of cilia on a wing-like protuberance. At some little understood signal, the veligers settle out of the water en masse, seeking a hard, clean surface upon which to settle. Those that do not find the proper surface die. Their ultimate preference, refined over the eons, is oyster shell. Oyster farmers know this and scatter broken shells (cultch) on

PLATE 23

A. Brown spiny sea star, *Echinaster spinulosus*, attacking an oyster, *Crassostrea virginica*. Reddish-brown with large tan or orange spines. To 6 inches across. Hundreds of the sea star's tube feet are simultaneously pulling on the oyster's valves (shells), but it is not necessary for the sea star to pull the shells apart to eat the soft inner tissues. Its stomach walls are so thin they can fit through imperfections in the shell's closure. The bright orange spot off-center on top (madreporite) allows water to enter and leave the water vascular system. The animal has no blood.

B. Underside of Forbes' sea star, *Asterias forbesi*, showing ambulacral grooves containing tube feet and a central mouth from which the sea star can extrude part of its diaphanous-walled stomach.

C. Clump of oysters showing typical inhabitants of an oyster reef. Mussels, brittle stars, barnacles, sea anemones, and fanworms are visible. The small crab is a species of *Pinnixa*, a white commensal crab that lives inside the oyster, plucking undigested food from its mantle edges. Sometimes it takes a bite from the oyster's mantle, becoming a parasite because it is harming its host. Oysters are clumped together, reducing feeding efficiency.

D. Enemies of the American oyster, *Crassostrea virginica*. The oyster is being attacked by an oyster drill, *Urosalpinx cinerea*. Using an acidic secretion, the snail perforates a hole in the shell and rasps out the oyster's tissues. It has glued clumps of flask-shaped eggs to the shell's surface. Two harmless jingle shells, *Anomia simplex*, are attached to the upper shell next to tubeworms. Small parasitic snails, *Odostomia*, are feeding on the oyster's tissues at the front of the lower shell.

E. Forbes' sea star, *Asterias forbesi*, regenerating two arms. Some sea stars reproduce primarily by fragmentation. The animal tears itself in two and each half regenerates a new body. One species, *Coscinasterias*, loses count and often produces offspring with six, seven, eight, or nine arms instead of the usual five. As long as each piece has part of the central disk, regeneration can occur. Here, two smaller arms are regenerating and will eventually become full-size arms.

PLATE 23

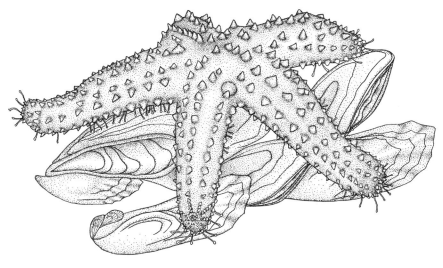

A.

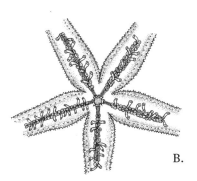

B.

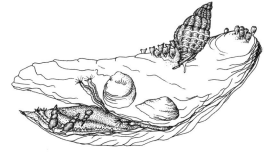

D.

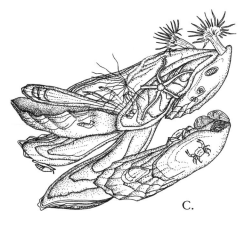

C.

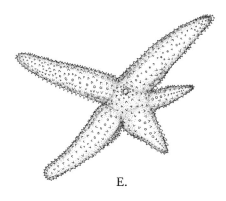

E.

the bottom for the veligers to settle on. But this leads to crowding, as several veligers settle on the same shell fragment. The oysters are distorted by their abutting brethren and suffer from competition, causing slow growth. Would it not be possible to fool the oysters into settling onto some artificial surface where their growth could be regulated?

My friend spent months trying to find the proper substrate that the settling larvae would like. He finally succeeded! Sheets of X-ray film dangling in the water were accepted. Furthermore, the flexible sheets could be bent and the now dime-size "spat" would pop off. The oyster farmer could take these now almost-impregnable juvenile oysters and strew them on the bottom so that they would not be crowded.

This important contribution was accomplished while my friend was working in the laboratory of a well-known oyster specialist. This man published a paper describing this research and *did not include my friend's name on the paper, denying him credit for months of research.* Although this is legal, it is unkind. No secure scientist does this. My friend became discouraged. He dropped out of school and became a llama farmer.

❧❧❧ Another Bad Guy

The grumpy, fuzzy, scholarly professor mentioned in chapter 21 had a son who joined the marines. When he returned from his tour of duty, he became a bayman, making a living harvesting wild clams on public bay bottoms. An unscrupulous leaseholder was using a modified commercial fishing vessel to vacuum clams from his sandy bottoms. When he had denuded his private leasehold, his boat "wandered" over the boundary to unfairly harvest clams on the public bottoms. Our former marine-cum-bayman saw this happen day after day, and finally, unable to stand by and watch his livelihood being taken away, leaped aboard the boat, muscular arms and tattoos flashing. The greedy captain locked himself in the cabin. Our hero proceeded to empty the sacks of clams into the bay. Upon his return, he was arrested. The captain had radioed ahead. Our impetuous bayman was eventually acquitted.

24

A Peek into the Anus
of a Sea Cucumber

THE SUN IS SETTING. The blasting intensity of the tropical sunlight dims. Shadows creep from the coral heads like black fingers grasping the grassy bottom. Darker shadows stir from grottos and niches in the coral. A nighttime population replaces the light-loving inhabitants of the reef and turtle grass bed. Soldierfishes and bigeyes, identified by their red color and large, light-absorbing eyes, drift slowly up from their daytime hideaways in the reef. The parrotfishes and damselfishes descend into the dark arms of the coral.

In the turtle grass meadow, all seems still. Then an almost imperceptible movement whispers through the shallow water. In this Alice-in-Wonderland tableau a foot-long, salami-shaped animal is placidly scraping sand into its mouth with twenty brushy tentacles, undisturbed by day turning into night. Its mouth full of sand, there is no room for life-supporting oxygen to enter, so another opening must do. At the opposite end, the multifunctional anus opens and closes regularly with each "breath." A sea cucumber is carrying on its life functions. The slow, rhythmic opening and closing of the anus reflects the sluggish metabolism of the animal. This remarkable species, the five-toothed sea cucumber, *Actinopyga agassizi*, unlike all others of its kind, has five square teeth ringing its anus. Why they exist there is one of the mysteries of the living world. They do not chew, they do not protect.

183

As the shadows darken, a pointed silvery head peers out from between the teeth. In total darkness a stealthy, slithering movement reveals the departure of a four-inch silvery pearlfish, *Carapus bermudensis*, sinuously searching the bottom for tiny organisms to pluck up with its narrow mouth. This fish must return to its coffin-like lair at dawn's first ray, like a Dracula of the sea.

૭૭૭૭

After a night of foraging, the fish finds an anus—any anus will do. It waits for the cucumber to exhale, thereby opening its anus to its full diameter. Then it darts into the anus, pointed tail first, and takes up its daytime occupancy of the sac-like chamber (cloaca) at the posterior of the cucumber's digestive tract. Evolution has provided the fish with a number of adaptations to its symbiotic existence. Its almost finless body lacks any impediment to entry into the anus—even the tail fin comes to a needle-like point, facilitating entry. Its digestive tract coils into a U-shape, permitting the anus to be located near the anterior of the body, so that when the fish must defecate it needs only to project partly from its rectal refuge. This adaptation provides safety while preventing fouling the fish's closet-like home. Formerly the pearlfish was considered benign, benefiting only from its refuge inside the cloaca. Recently, it has been upgraded to the level of parasite, since it has been observed to tear mouthfuls of internal organs from its host. It is thus a rarity, a vertebrate parasite of an invertebrate host.

૭૭૭૭ Sea Cucumber Salad

Like a cow, the sea cucumber spends the day munching on its cud (or in this case, its mud). To a greater or lesser extent even the most pristine sand contains minute particles of organic material in the interstices between grains.

Something dies. It disintegrates into tiny pieces of once-living tissue (detritus). These settle into the sediment, fragments of the dead. They accumulate on all surfaces under the sea. The sand seems to absorb them like a sponge. The sea cucumber extracts this once-living material to incorporate the molecules into its own tissues. It rearranges these molecules of dead plants and animals into sea cucumber protoplasm.

Twenty-four hours each day, the animal sweeps sediment into its mouth. Peristalsis, the rhythmic contraction of the gut, forces the gritty sand backward, inevitably irritating the intestinal walls, causing the copious secretion of mucus. The sand is flooded with enzymes. Each enzyme is specific. One breaks down plant starch into sugars, providing energy-laden fuel. Another enzyme tears animal tissue into its constituent amino acids. Once these molecules pass through the walls of the gut, they are reconstituted into a different arrangement—sea cucumber protein. The sand passes further down the gut, now bereft of all organic constituents. With a final constriction, the rectum releases its burden of immaculate sand, now nicely wrapped into a hollow, mucous-covered cylinder, the fecal cast.

The fecal casts allow the initiated to identify the rear from the front of a supine sea cucumber. The animal is so sluggish that the fecal casts, discarded remnants of yesterday's sandy dinner, lie near the anus all day. The sophisticated snorkeler, observing the casts near the cucumber, thinks, "Ah hah! That must be the posterior end." This valuable piece of knowledge separates him or her from the ignorant boobs snorkeling nearby.

ꜱꜱꜱꜱ Magnificent Anus

The multifunctional anus has, of necessity (the mouth being perpetually full of sand), assumed the functions normally performed by the mouth—"breathing" and defense.

Relaxation of the muscles constituting the body wall enlarges the body cavity, creating a "vacuum." Oxygenated water is sucked into the anus with force enough to distend a branched respiratory tree. Each tiny branch is tipped with a bunch of "grapes," alveoli-like thin-walled spheres, like a

PLATE 24

A. Donkey dung sea cucumber, *Holothuria mexicana*, dark brown above, bright rose or white below with brown podia (tube feet) scattered on its bottom. To 16 inches. This one is placidly shoveling sand into its mouth with twenty brush-like tentacles in a turtle grass bed. The sandy tubes are fecal casts released from its anus.

B. Sea cucumber under stress spewing out its internal organs. When attacked or placed in an adverse situation, the animal eviscerates some of its internal organs through its anus, distracting the predator. The internal organs can include part of the respiratory tree and intestine. The animal eventually regenerates these organs. Some sea cucumbers expel brightly colored innards; some eviscerate poisonous organs. Toxicity is determined by placing small fishes in an aquarium containing the sea cucumber. The time interval between evisceration and the death of the fishes is calibrated as the degree of toxicity.

C. Front of a sea cucumber showing mouth surrounded by twenty peltate (brush-like) tentacles for sweeping sand into the mouth.

D. Pearlfish, *Carapus bermudensis*, appears to lack fins. The tail is pointed for an easy entry into the five-toothed sea cucumber's anus. Silvery, to 6 inches. The pearlfish's digestive tract curls around into a *U* and its anus opens just behind its gills so it can defecate without extending its full body length out of the sea cucumber.

E. Pearlfish entering the anus of a five-toothed sea cucumber, *Actinopyga agassizi*, tail first after a night's foraging. Any sea cucumber anus will do, but despite its teeth, *Actinopyga* is very much favored. When the fish touches the anus, the sphincter closes tightly. The fish waits until the sea cucumber exhales, relaxing its sphincter, and backs in.

PLATE 24

A.

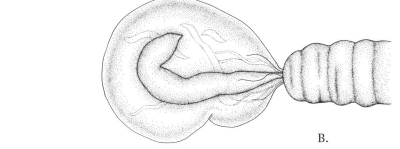

B.

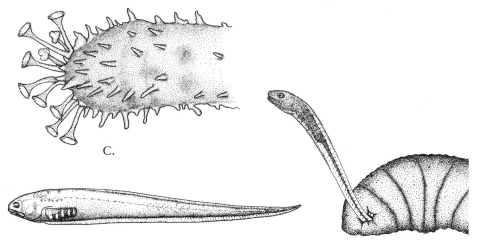

C.

D.

E.

bunch of balloons. The water, squeezed into these balloons, gives up its oxygen, which diffuses throughout the whole body cavity. Periodically the muscular body wall constricts, expelling water from the respiratory tree in a perceptible flow out the anus.

But it is the defense function that is so unusual. One might think of a threatened animal rearing up, opening its mouth wide, baring its fangs. But the only teeth *Actinopyga* has are in its anus, and besides, its mouth is full of sand. All other sea cucumbers do not have anal teeth. They present an even more innocuous defensive appearance. Are these slug-like animals vulnerable? Surprisingly not. With fantastic evolutionary imagination, sea cucumbers have become masters of defense.

A predator attacks. The cucumber's first line of defense is a thick body wall studded with tiny calcareous spicules called ossicles, a veritable chain mail armor. (These have two characteristic shapes: "buttons" and "tables.") The predator drops its prey, unable to penetrate the tough, almost impregnable body wall.

Suppose a determined predator picks up the cucumber, shaking it about with fierce determination, teeth grinding. Defense mechanism number two is triggered: with a powerful "blurch," colorful internal organs are ejected from the anus! The predator is distracted by the tasty innards and the cucumber slowly creeps away on thousands of tiny tube feet, to hide somewhere and regenerate new internal organs. In some cases the eviscerated organs are poisonous and the predator gets a mouthful of toxin. It will not attack a sea cucumber again.

But suppose the predator is not fooled and returns to the attack. It is met with a hail of sticky arrows, Cuvier's tubules. Some cucumbers will readily expel these last-ditch defensive darts. One, *Holothuria impatiens*, is named for its tendency to impatiently release them even when mildly disturbed. Once, while surveying cucumber populations on the Great Barrier Reef, an innocuous-looking specimen released tubules that crisscrossed my arms in a white spider's web. I tried to remove them. It felt like tearing off a long, tenacious Band-Aid. A juvenile cucumber can trap a predaceous crab in a web of sticky tubules, but an adult is too large to be attacked by anything below the size of a stingray, rendering the tubules virtually useless. Can it be that Cuvier's tubules are retained juvenile characteristics? Such a sticky question cannot be resolved in the fossil record because the tubules are not hard and calcareous, and will rot before fossilization can occur.

One of the only animals able to pierce the almost invulnerable armor to gain sustenance from the cucumber is a tiny, sharply pointed snail, *Melanella*. This parasite has forsaken snaildom's traditional file-like tongue and uses enzymes to eat away the body wall. It then inserts a long proboscis into the cucumber's body cavity and sucks out the nutritional coelomic fluid. Four or five of these minuscule, white, shiny snails can often be seen attached to the undersides of sea cucumbers.

The only predator that can always bypass the sea cucumber's extraordinary defense system is the most terrible of all—humans. Pacific Islanders collect armfuls of cucumbers and place them in sun-drenched tide pools. The water temperature rises to over 100°F. When stressed, the cucumber cannot employ a defense mechanism appropriate to the challenge. It cannot vary its instinctive response. It has no alternative but to employ its inborn defense mechanisms. It eviscerates its internal organs. This self-cleaning mechanism is a boon to the hungry harvester. He or she picks up the now-limp cucumber and rubs it vigorously against sharp seaside rocks, shearing off the skin in which the tiny ossicles are embedded. The gray tubular animal is then sliced and added to soup, the thick, soft, gray circles resembling mushrooms. The Pacific Islanders call sea cucumber dishes *beche de mer* or *trepang*. In Mandarin Chinese it is called *hoy tam*.

Those "mushrooms" you ate when you last had soup in a Chinese restaurant may have been slices of sea cucumber.

25

The Yellow Submarine

THE LAUNCH seemed to be circling aimlessly. Suddenly, as if in a movie, the bow of a submarine broke the water about twenty feet from us. It was a far cry from *The Hunt for Red October*, and the chubby man who poked his head out of the conning tower was no Sean Connery.

The submarine was tiny, no more than twenty feet long, and its bow was a five-foot-diameter convex glass lens through which could be seen two raunchy-looking characters in faded tan shorts. This was a research vessel and the two oddballs were well-known marine scientists. We pulled up alongside and leaped aboard the precariously curved deck of the sub. After the usual banter, crews were exchanged. We slid down inside the conning tower and took our positions in the bow, peering out at a 180-degree view of the boundary layer between the sky, and the clear tropical water.

Adjustments were made by the captain. He turned on the carbon dioxide-absorption apparatus and opened the oxygen valve. "OK?" he asked. We nodded our assent and he began to flood the ballast tanks. We descended slowly, a gentle bump signifying that we reached the bottom at sixty feet. There, almost touching the window, was a six-foot, very disturbed green moray, *Gymnothorax funebris*. It is the *funebris* part of the name that gets me every time. I don't mind the fact that the slimy green skin of this beast is covered with bacteria-laden mucus, and that its huge mouth is studded with backward-facing pointed teeth—it is the funereal species name that bothers me. It glared at us. I glared back, glad to be separated from it

190

by inch-thick glass. Visibility was uncanny through the five-foot, optically-corrected lens that was the bow. It was like looking through air.

The boat moved horizontally over the coral reef, which glowed with vibrant colors as the sun shone down through sixty feet of water. Brightly colored fish sparkled and shimmered. Schools of large predators circled black clouds of silverside minnows, which frantically squeezed together lest one be singled out for attack. The reef was resplendent in all its glorious color, spread out before us in the transparent water.

Our descent continued. First, as far as the eye could see, the bottom graded to a field of antler-like staghorn coral, *Acropora cervicornis*. Then the slope sharpened; the antlers became sparse, separated by larger and larger sandy areas. As the light became more tenuous, fewer corals appeared. Huge, isolated bommies protruded ten feet from the sand. These gigantic heads of boulder coral, *Montastrea annularis*, were a haven for sponges and other corals that thickly covered their vertical sides. The hangers-on were safe from suffocation in the increasingly prevalent soft sand. Purple and yellow six-foot-high feathers and fingers of soft corals formed lush gardens protruding from the sediment.

Then the bottom began to slope ever more sharply toward the abyss. Sand was everywhere. Corals declined in numbers and species. Their place was taken by gigantic sponges. Some thrust hollow yellow tubes towards the surface. Bumpy red balls of sponge lay strewn about. Brown crusty sponges covered the bottom in yard-wide patches. By far the largest sponge was the bathtub sponge, *Xestospongia muta*. This species has been made famous by jocks. This term describes scuba divers whose sole interest is in demonstrating their masculinity (even female divers). They do not look around them at the startlingly exquisite unreality of the coral reef. Their goal is to dive deeper than the next man (or woman). Somehow they have heard that the bathtub sponge usually appears in water deeper than one hundred feet. The objective is for the diver to sit in one of the "bathtubs" and have someone take a picture of him or her with an arm raised, as if washing an armpit.

The bottom began grading down to the deep fore reef. One hundred feet . . . two hundred feet. The light was becoming attenuated. Finally, the sloping terrace sharply descended to a cliff, beyond which lay the blackness of the abyss. We were approaching three hundred feet. The edge was visible. Only the mysterious darkness of the unknown depths lay ahead. Our mood became somber as the whining sound of the electric motors turning the propellers became louder as the attitude of the vessel changed, the angle of descent becoming sharper. The captain pointed our bow window towards the almost vertical wall and kept us about three feet away. We slowly descended into the seemingly unending depths.

At three hundred feet the wall seemed to be soft ooze penetrated by mysterious holes. Sprigs of black coral, *Stichopathes*, jutted from the steeply angled slope, interspersed with a few unfamiliar species of soft coral. The captain turned on the lights and the soft bluish haze was lifted. The black corals were made fuzzy by now-visible extended polyps. The brownish soft corals became red as the normal light spectrum was reinstated. We descended further.

Suddenly, in a little depression I saw what few men have seen. My heart raced and I gasped. There in front of me was what appeared to be a woman's breast. I do not normally gasp at such a sight, but I immediately recognized this object as the rarest of finds. It was the rounded mass, complete with "nipple," characteristic of *Ceratoporella*, a member of an obscure new class of sponges, Sclerospongia. Surprisingly, this class was first described recently (in 1970). It was thought that all the major taxonomic categories were established by the end of the nineteenth century. Few taxa as encompassing as a class have been described in over a hundred years. But this was a recent exception, and was celebrated by the few of us who realize its importance. I never thought I would see a live sclerosponge. *Ceratoporella* is rarely found above two hundred feet, well beyond the scuba diving capacity of all but a few brave souls. Its discoverer found a specimen in a cave close to maximum diving depth. We were seeing this sponge class out in the open, perhaps one of man's first such viewings.

We descended to four hundred feet. The only sound was the muted hum of the electric motors. This time it was the captain's turn to get excited. He had noticed something that stood out from the soft, muted, slaty color of the wall. It was camouflaged and I had not noticed the slit shell snail, *Entemnotrochus adansonianus adansonianus*, a deep water species. At the lab, workers were culturing a small group of these snails, and this specimen would be welcome. Surprisingly, they were surviving in an aquarium although their normal habitat was the soft sediment and high pressure of the abyssal depths. The vessel swung around and the mechanical arm was extended. With great dexterity the captain maneuvered the arm so that its "hand," three moveable finger-like steel digits, grasped the fist-sized snail and dropped it onto a mesh tray directly below us.

The slow descent continued . . . five hundred feet . . . six hundred feet. The wall became a slate-gray desert, only a few mysterious holes and sinuous tracks marking the featureless slope. Occasionally we came too close and disturbed the superficial silty surface, producing a puff of fine sediment, defiling the ancient smoothness of a velvety layer unsullied over thousands of years. It was composed of the calcareous and siliceous skeletons of countless microscopic dead planktonic organisms that had been raining down for eons.

The whining of the motor slowed and we felt a bump. "Seven hundred forty feet," announced the captain. He shut off the power. The lights went out, the whining stopped. We were alone in inner space. We heard the silence of infinity.

As our eyes adjusted, I noticed a barely discernable bluish aura out the window. It is commonly thought that visible light disappears at the boundary of the photic zone, the region where light is sufficient to support photosynthesis. It is reported to be about 300 feet deep. I peered out. I could see something! Surprised, I realized that dim blue light had filtered through 740 feet of transparent water. Then in a moment of epiphany I understood that light penetration in the sea varies with the interference of light-scattering planktonic populations. In the clearest of tropical seas, virtually devoid of plankton, light penetrates farthest—even to more than double the traditionally quoted depth. Of course, the filtered bluish light was so feeble as to be unable to support photosynthesis. This is truly the twilight zone. Any life on the bottom must subsist on droppings from above—dead animals and plants whose lives were lived out in the sunlight-penetrated layer.

I heard the captain musing to himself. "Wait a minute," he muttered. He was looking out of the window as if he was on 42nd street and searching for Broadway. He grunted with satisfaction and turned us in a specific direction. The lights were turned on. They revealed just a few silt-covered humps and some recent, uncovered coral fragments. The flattish deep-water wasteland stretched out to the penumbra of glaring light coming from both front headlights. After about ten minutes of moving at maximum speed of three miles per hour, he grunted, "There it is." Before us lay the supreme, the ultimate—the object of our quest. On a rock jutting a few feet from the silty bottom was a virtual garden of *five stalked crinoids. Cenochrinus asterius.*

A crinoid is an upside-down sea star. Whereas a sea star crawls over the bottom and eats buried clams, its relative, the crinoid, is rooted to the bottom attached by a stalk, but facing upward to catch plankton. All echinoderms are derived from an ancestor that was radially symmetrical, with five arms or multiples thereof, extending in all directions. The sea star, unattached, can crawl in every direction with equal ease, threatening clams and oysters anywhere. But crinoids are sessile (attached) and are vestiges of an era, 450 million years ago, when few ancient sea animals had barely evolved cephalization, the accumulation of the senses at the front of an organism. A front means that ears, nose, and eyes can be coordinated with mouth, and prey can be chased and engulfed. In the soupy seas that existed before cephalization evolved, it was advantageous to stretch out many plankton-collecting arms in every direction, the easier to collect a meal. But gradually, the more efficient animals with anterior sensory specialization reduced the lavish populations of plankton.

Before the dominance of cephalization, the crinoid was king. Shale beds, muddy bottoms of ancient seas turned into rock, are literally covered with dime-size circles, the fossilized holdfasts of stalked crinoids. But the era of these stationary stalked crinoids was over and they became extinct—all but a small population that remains to this day in the deepest, coldest, blackest recesses of the abyss.

৩৩৩৩ The Scariest Crinoid

Ever since I was traumatized, I have never been a confident diver (see chapter 5). I do it because it is part of the job. One day, off Bonaire, I found myself moving down a sloping wall on my own—without the enveloping walls of a submarine. My diving technique is to carefully examine the downward angling bottom, keeping close, and never look up through the shoals of iridescent blue chromis, *Chromis cyanea*, to the silvery surface that lies what seems to be an infinite distance away. But there is a disadvantage to this approach—one loses touch with the surface. I dove ever deeper, taking care to never look up. When I reached the sand at the bottom of the slope, I looked up and was horrified. The surface seemed to be an endless distance away. I found out later that the sand started at 110 feet!

Keeping close to the upward sloping wall, I made a slow ascent, forcing myself to examine every nook and cranny of the reef and not look up. About halfway up, I saw a wonderful sight. There, perched on a coral head, was *Nemaster grandis*, the grandest of the modern-day crinoids. About 200 thousand years ago, a crucial evolutionary step was taken. A new suborder, Commatulida, branched off from the ancestral stalked crinoids. It had evolved foot-like appendages and had a newfound mobility that led to great success. Many new species evolved. Some could swim! They were able to seek out plankton-carrying currents and could hide. Tiny sensory cells embedded in the epidermis allowed the commatulid crinoids to seek out the darkest havens. Only a gold or orange arm draped over the coral surface revealed their presence. But *Nemaster grandis* is different. It displays itself in all its glory. Forty fifteen-inch, comb-like black and white arms project into the water column, each straining out plankton that is then carried down a food groove in each arm to a central mouth.

Fascinated, I approached for a better look. Then I noticed that *my hand was creeping forward to touch the crinoid*. I tried to will my hand back, but it seemed to have a mind of its own. Slowly it inched forward—then a finger touched the bottom of the animal. Instantly the arms, which had been suspended in the water column, snapped down on my hand! As I jerked it free, a profusion of bubbles burst from my regulator, each signifying fear. After closing my eyes and hanging suspended in the warm, enveloping water, the panic subsided and I slowly ascended the remaining fifty feet to the surface.

୬୬୬ The Yellow Submarine

There is a remarkable building on the shore of the city of Elat in Israel. A white rotunda lies about a hundred feet offshore, connected to a complex of buildings housing a public aquarium. One walks along a catwalk to reach

PLATE 25

A. BLACK-AND-WHITE FEATHER STAR, *Nemaster grandis*, flamboyantly displaying forty black-and-white arms on a pillar coral. This commatulid crinoid can crawl; some commatulids can swim. An extraordinary step up from sessile (attached) sea lilies, the evolution of movement in crinoids led to an efflorescence of species in the last 200,000 years.

B. ADANSON'S SLIT SNAIL, *Entemnotrochus a. adansonianus*. Tan with rose or purple bands, shell grooved, 6 whorls. A distinct slit is located above the first whorl. To 3 inches tall. Of 1,500 named species, only 30 are found today. Live on walls in deep water, from 100 to 2,500 feet. Eats sponges and other sessile invertebrates.

C. GOREAU'S SCLEROSPONGE, *Ceratoporella nicholsoni*, yellow, tan, or brown, 3 inches to 3 feet in diameter. Smaller sponges are lumpy, round, with nipple-like protuberances. Grows slowly, inches per century. For this reason it is used to measure the salinity and temperature changes of the Caribbean over time. Large specimens may be 1,000 years old. Found on the slope of coral reefs below 150 feet or in caves at shallower depths.

D. GREEN MORAY, *Gymnothorax funebris*, is bright green. To 5 feet. One stared at us through the bow lens of a submarine at sixty feet. Sharp inward-pointing teeth prevent prey from escaping. Usually hidden in a coral crevice. Bites only when interfered with or when a hand is inserted into its lair. Its blue skin is covered with yellow mucus, producing a green color.

E. CARIBBEAN SEA LILY, *Cenocrinus* species. White, pale tan, to 14 inches with about 16 whorls of flexible projections (cirri) on stalk. Five *Cenocrinus* on a rock 740 feet below the surface, as seen from a submarine.

Usually found deeper than 500 feet on hard objects projecting from the sea floor or on walls. Attached to the bottom by root-like holdfasts. Sea lilies (stalked crinoids) were most populous 400 million years ago, dominating the early fossil record. Fewer than eighty species exist today.

PLATE 25

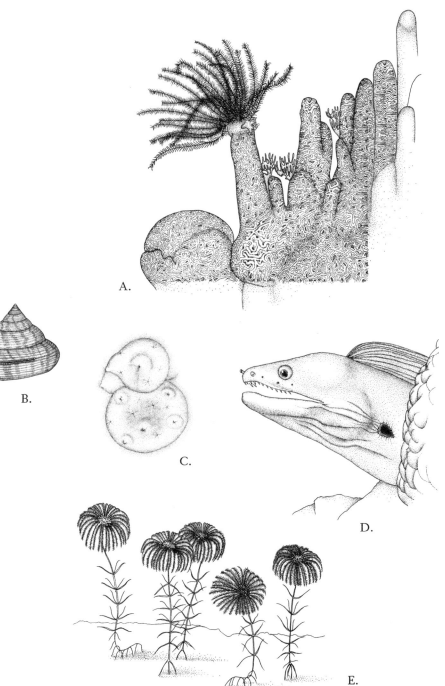

A.

B.

C.

D.

E.

the semi-submerged rotunda and then descends stairs to reach an underwater world. There, a lush coral reef supports a veritable horde of mind-boggling Red Sea fishes and invertebrates disporting themselves in front of glass windows encircling the building in a remarkable undersea vista.

A magnificent lionfish, *Pterois volitans*, swims purposefully toward the window, undulating its flamboyant fins, warning predators away with each of its poison-tipped spines. Bright red gobies dart around a pillar of coral in a gorgeous, circular mating ritual. Clouds of goldfish-like jeweled fairy basslets, *Pseudanthias squamipinnis*, rise from holes in the coral. The cloud descends into the holes upon the approach of a predator, leaving a vacuum of color until it passes. Then, as if assured of safety, the orange cloud rises—hundreds of fishes ascending inches above the coral to resume feeding. Huge beetle-browed supermale parrotfishes munch on coral heads, defecating a cloud of white calcareous powder—the undigestible remains of their last meal. A spotted unicornfish, *Naso unicornis*, its single "horn" projecting several inches from its head, lines up to be cleaned by schools of cleaner wrasses, *Labroides dimidiatus*. A foot-long spanish dancer nudibranch, *Hexabranchus sanguineus*, sinuously twirls through the water, voluptuously flashing its red and black petticoat-like mantle walls.

The scene shifts from vista to vista as one walks around the large submerged room. Always the exquisite beauty of the coral reef permeates one's consciousness. It is remarkable that this reef is flourishing so healthily when the corals in the rest of the world are dying. A thought comes to mind, "After all, this is the land of miracles."

It is not completely satisfying to peer through a window at the top layer of a reef that extends down for several hundred feet. What must it be like to see the deeper parts? Will there be crinoids? Huge 400-pound jewfish, *Ephinelus tauvina** (where else if not here?)? Sharks? Sea turtles? Huge sea cucumbers? The chance to go more deeply into the subject lies at a berth nearby. At the end of a pier near the aquarium is moored the proverbial Yellow Submarine. Built in Belgium, this bright yellow vessel looks more like a bus than a sub. It is designed to take tourists on underwater tours, just like the buses outside take visitors to ancient cities in the desert. Upon entry

---

* Any number of large groupers are named "jewfish." For example, *Epenephelus itajara*, *Garrupa nigrita*, and *Stereolepis gigas*.

the stewardess directs you to your seat and asks you to fasten your seat belt. There are about twenty seats along each side, each with a window. A muted bell rings and the vessel begins to move seaward. Another bell sounds, we begin our descent. I peer out through the transparent water. We reach one hundred feet. Everything outside appears bluish.

To my astonishment, I am looking at a veritable wasteland—dead coral everywhere. Each colony lies neatly in place. It is as if a horde of barbarians has laid waste to an ancient city. (After the attack, houses become time capsules—dishes on the table, wine in the decanter, shoes under the bed. When excavated, archaeologists are able to peer backwards with the aid of these fixed moments.) So it is with the coral reef. This richly diverse Red Sea coral reef (there are about ten times the number of coral species in the Red Sea compared to the Caribbean) is dying. Blue-gray skeletons are all that is left. Gone are the dominant lettuce-like colonies, the huge coral boulders, the coral antlers. Gone are the fishes, worms, and snails that lived out their lives in the protective nooks and crannies of the reef. Gone is the rich ecosystem based on the extraordinary coral animals, able to survive in the clear cerulean seas. Already silt is covering them.

We descend to about two hundred feet. Nowhere is any life visible. How is it that we saw a profoundly profuse population of animals from the windows of the rotunda, while death surrounds us in the submarine? We begin to ascend. The tour is over and we head for our pier. It is then I see the answer. The underside of the rotunda is clearly visible about fifty feet away. Around it is a huge wire mesh "tray." The designers of the underwater building had built a platform around it and filled it with chunks of living coral and populations of animals. What we saw was an artificial living exhibit of what had been.

This reef is as near death as all the others on the earth. The optimism of the Beatles was a reflection of a different era, one of flamboyant coral reefs—no more "living beneath the waves in our yellow submarine."

26

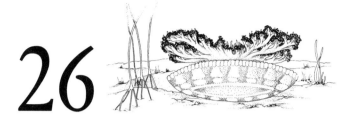

The Perils of Vanity

ONCE UPON A TIME there lived a beautiful princess named Andromeda. Her father was the rich King Cepheus and her mother was the renowned beauty Queen Cassiopeia. Unfortunately, the Queen knew how beautiful she was, and in a more assertive manner than the witch in Snow White, looked at her reflection in the sea and said, "I am the most beautiful of all." She was overheard by the ubiquitous sea nymphs called nereids. Each nereid, being partly immortal, fancied herself the most beautiful of all, and took great offense. They complained to their daddy, Poseidon, who decided to lay waste to the land and not stop until Cassiopeia's only daughter, Andromeda, would be sacrificed to him. Sadly, Cassiopeia and Cepheus tied their daughter to a rock in the sea. Poseidon untied his pet sea monster, telling him to have his way with Andromeda.

Coincidentally, the hero Perseus happened to be flying by on Hermes's winged boots and looked down at the scene. He said to Cassiopiea and Cepheus, "I am Perseus, son of Zeus and Danae. I slew the evil monster, Medusa. I can fly. I am some catch as a son-in-law. I hereby ask for permission to woo your daughter." Who could resist those credentials?— especially since the sea monster was nearing. The king and queen said, "We will be happy to have you as our son-in-law, and we will throw in our kingdom as a dowry."

Perseus slew the monster and the happy couple were united.

But . . . King Cepheus had a brother, King Phineus, who had been betrothed to Andromeda before her unfortunate bondage on the rock. He ran away when the sea monster approached, but when Perseus and Andromeda were preparing to marry, he returned with his army to claim Andromeda. Everyone was amazed because he had forsaken Andromeda with his cowardice. Now he claimed her, ignoring the fact that Perseus was her savior.

His army attacked. Perseus was hard pressed and outnumbered. Then he remembered his ace in the hole. He reached into his knapsack and took out the severed, snake-crowned head of Medusa, the Gorgon. Anyone who looked at the head would turn to stone. He angled his reflective bronze shield so that the Gorgon's face glowered at his enemies. Instantly he faced not a ferocious army, but a group of stone statues. To this day, all jellyfish are called medusae because any small animal that comes into contact with their snake-like tentacles is paralyzed, as if turned to stone, by microscopic poison darts called nematocysts that cover each tentacle.

After living a long and full life, Cassiopeia died, for she was half mortal. The gods debated her fate after death. The nereids never forgave her vanity. They campaigned against her and Poseidon acquiesced to their wishes. She was made into a constellation that is located high in the northern sky, upside-down for six months of the year, an ignominious position for such an illustrious queen.*

᭥᭥᭥᭥ Cassiopeia's Biological Legacy

The upside-down jellyfish, *Cassiopeia xamachana*, is named after the queen because of its habit of flopping over to expose its undersides to the sun's rays. It lies upside down on the mud, pulsating gently, undulating its petticoat-like lappets toward the sun. Why does it perform this ritual? Surprisingly, it rarely eats zooplankton like other jellyfish. *Cassiopeia* obtains its energy from foreign invaders—the photosynthetic activity of millions

---

* Should you be interested in the constellations Cepheus, Perseus, Andromeda, and Cassiopeia are in the same northern quadrant.

PLATE 26

A. Upside-down jellyfish, *Cassiopeia xamachana*, bathing its zooxanthellae (mutualistic one-celled dinoflagellates) in sunlight. The jellyfish's transparent tissues are filled with zooxanthellae, giving them a golden-brown color. Capable of carrying on photosynthesis, zooxanthellae produce food that is the jellyfish's primary source of nutrition. It is likely that zooxanthellae were once free-living microalgae that became incorporated into tissues of jellyfish hundreds of millions of years ago. This is a logical hypothesis because there are many species of free-living dinoflagellates that carry on photosynthesis.

Cassiopeia is usually found on muddy bottoms among red mangrove trees, gently pulsating. The bell is brown and white. To 12 inches in diameter. The mouth is surrounded by large extensions, oral arms, that are edged with frills called lappets. The animal may eat some zooplankton, providing needed protein.

This is Cassiopeia the animal.

B. Cassiopeia, half mortal and half goddess, was the mother of Andromeda and the wife of King Cepheus. She bragged about her beauty while looking into the smooth, reflective sea. Her vanity offended the sea nymphs collectively called nerieds, each of whom thought she was the most beautiful of all. At their request Poseidon, their father, arranged for her memorial constellation to be located so far in the northern sky as to be upside-down for half the year, a most ignominious position for a queen. Here Cassiopeia is seated on her throne holding a feather.

This is Cassiopeia the myth.

C. Constellations of the northern sky. The compass rose above the constellations and to the left of the mythical Cassiopeia indicates due north. During the wintry half of the year, Cassiopeia is in the northern quadrant and upside down.

   i. Cepheus

  ii. Cassiopeia

 iii. Andromeda

This is Cassiopeia the constellation

PLATE 26

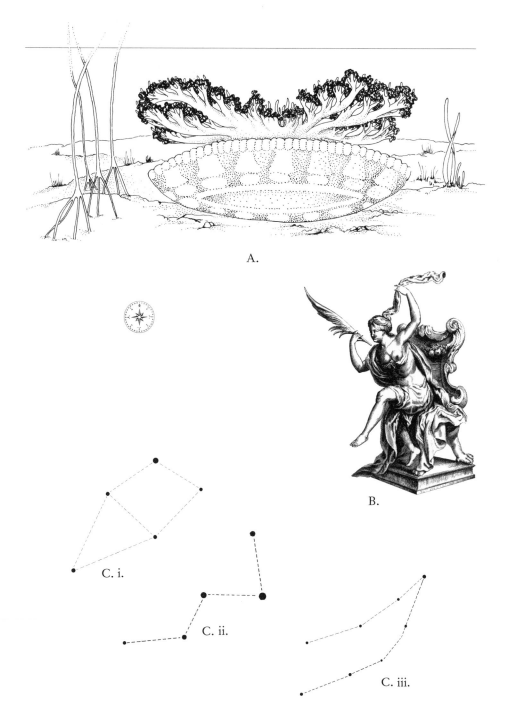

A.

B.

C. i.

C. ii.

C. iii.

of plant-like (algal) cells that have occupied its tissues. These cells, called zooxanthellae, live in a mutualistic relationship with the jellyfish. Nestled in the protective host tissues, the single-celled zooxanthellae use carbon dioxide and water, wastes produced by the respiration of the jellyfish, as raw materials to produce their own food, and release enough to feed their host in an intimate trophic relationship. Except for a consuming few zooplankters to provide protein, these jellyfish are self sufficient. In muddy areas in the tropics, you can see hundreds of these foot-wide upside-down jellyfish, gently pulsating, bathing their zooxanthellae in a filmy ballet.

～～～ Coral: Part Animal, Part Plant

Zooxanthellae make it possible for abundant life to survive in desert-like tropical seas. Why are tropical seas their transparent azure or turquoise color? Because the reflection of the blue sky is not obscured by thickly growing green phytoplankton, characteristic of higher latitude, well-nourished seas. The exquisite tropical waters we yearn to visit are, in reality, barren and impoverished. Only animals that do not rely on a food web dependent on free-floating plankton can survive here.

The base of the shallow tropical oceanic trophic pyramid is not plants, but "plant"-bearing polypoid cnidarians—corals. Indeed, if it were not for huge coral reefs (the Great Barrier Reef extends over 1,200 miles) the gorgeous underwater tropical vistas we think of would not exist. Sunlight penetrating the clear seas makes it possible for zooxanthellae to perform photosynthesis. Buried in the tissues of the coral polyps, they provide them with the energy to remove calcium carbonate from the warm water to build gigantic reefs, whose innumerable interstices are havens for the vast majority of living things in tropical seas.

The evolution of the coral polyp-zooxanthellae mutualistic relationship is a conundrum, the solution of which is lost in time. There is little doubt that a chlorophyll-bearing single-celled algal species took up residence in the tissues of coral polyps—there exist today similar free-living cells called dinoflagellates. But how did the zooxanthellae enter into such an intricate relationship? Living in close proximity over almost limitless time resolves all problems of symbiosis.* Nevertheless, it is hard to comprehend how the zooxanthellae so tightly tied their metabolism to the host polyp, and vice versa.

Sadly, the astonishing interaction between coral and zooxanthellae is being reversed by modern man. Global warming *does* exist. Proof lies in the sensitive nature of coral reefs. Having evolved in the constant temperatures of the open sea, corals are unable to stand even brief exposure to fluctuating temperatures. Present-day oceans average 2°C above past normal temperature. Corals, needing constancy, do not adapt. In their death throes, they expel their zooxanthellae. This sensitivity makes corals the canaries in the coal mines. They warn us of the impending doom of many of the earth's less flexible organisms. The life-sustaining zooxanthellae are being expelled from corals in all the world's oceans. The corals, bereft of their internal providers, die. The condition is called bleaching. In all the world's oceans the gold and green color of many coral reefs is gone, leaving a slimy white coating of dying polyps.

Be forewarned.

* Symbiosis means living together. Mutualism, commensalism, and parasitism are all subsumed under the heading "symbiosis."

27

Sexually Repressed Victorian Taxonomists

I HAVE AN IMAGE of Victorian taxonomists as cobweb-covered fellows locked in a museum attic, thinking sexually repressed thoughts. This image derives from their peculiar penchant for naming the most unappealing organisms after the most beautiful mythological goddesses. The most blatant example of these wandering thoughts would be the origin of the names of annelids (segmented worms). These phallic animals should ordinarily inspire one to think of male gods, like Zeus and Apollo, but since many eighteenth- and nineteenth-century taxonomists were frustrated males, we find a variety of wormy species with female nomenclature.

✿✿✿✿

Like these men, I find exquisite beauty in clamworms. As a group, they are called nereids, named after the lovely, ephemeral sea nymphs. But to use the name *Nereis** to refer to one of the most vicious of predatory worms seems

---

* The brother of one of my former students produced a successful Hollywood horror

excessive. She fiercely extrudes her fleshy pharynx to capture small worms, everting it explosively to plunge two black mini-fangs into her prey. The pharynx retracts and the unfortunate victim disappears into its fleshy folds. The mythological nereids, for all their beauty, were also dangerous—to offend them meant risking the wrath of their father, Poseidon.

ꕔꕔꕔ Godly Gossip about Willful Worms

The goddess Amphitrite, like a Gorgon, has a coiffure of writhing snake-like tentacles. In her namesake worm's case, they are used to gather detritus from the mud. She is the daughter of the sea god, Nereus, and one of his dalliances, the kindly goddess Doris. Mom is immortalized as a shell-less snail, a nudibranch, which resembles a lump of flesh with a tuft of gills extending from its posterior. Dad is the father of the nereids by another relationship.

Another lump is *Aphrodita*, the sea mouse—a fuzzy, mound-like, scaled segmented worm that, perversely, is named after the goddess of love. She was created out of an upsurge of frothing foam that exploded from the sea when the sun god, Cronus, cast the mutilated penis of his father, Uranus, into the ocean. The lustful Aphrodite had a dangerous dalliance with clever Hermes, god of thieves and scientists, producing sons Eros (Cupid) and Hermaphroditus, the original hermaphrodite, who made love with a sea nymph so vigorously that they fused together. Another relationship yielded Priapus, the fertility god, usually depicted with a huge erect penis. A contrarian Victorian taxonomist, accepting the masculine, phallic nature of worms, named an obscure wormy phylum "Priapulida" perhaps because of its species' tendency to become turgid and elongate when stimulated.

Aphrodite, the Greek goddess of love, evolved into Venus, the Roman goddess of love. In this latter incarnation, she inspired one of our frustrated taxonomists to name a *clam* after her. The commercial edible clam, the

---

film in which the clamworm, *Nereis*, enlarged to King Kong size, terrorized the countryside. The movie was called *Squirm*.

quahog, was formerly named *Venus mercenaria* before the name was changed, possibly out of embarrassment, to *Mercenaria mercenaria*.* This has inspired the bumper sticker, Eat Oysters, Live Longer; Eat Clams, Love Longer. In truth, the front of the clam has inscribed on it a perfect heart, the possible association with Venus.

The dangerous sea nymphs *Hermodice* and her sister *Chloea*, the fireworms, are strikingly beautiful—olive, red-bordered bodies edged with glistening white margins. To touch them is to be cursed, for these margins are composed of thousands of barbed and venomous siliceous mini-needles called setae. To pick up one of these fat, juicy, worms is to risk weeks of inflamed fingers as the barbed setae penetrate beneath your epidermis. Even fish have learned to avoid eating these tempting worms, which so conspicuously crawl about in turtle grass beds.

Hesione has long flexible projections extending from her heart-shaped head, and a crown of long, tentacle-like cirri. But don't be fooled by heart and crown. Four black eyes scour her surroundings as she waits for prey, which are trapped in the cirri and engulfed. This worm is misnamed, for its rapaceous nature does not reflect the innocuous charm of its namesake, who was chained to a rock, awaiting "in speechless terror" the arrival of a sea monster loosed by the ever-vengeful Poseidon. In this story the hero, Heracles (Hercules) happens on the scene. He leaps into the gaping maw of the monster and cuts out its entrails, slaying it. But would you think her father would honor his promise to Heracles? No, he does not offer Hesione in marriage to the hero, who sails off in a huff.

Some taxonomists also thought of jellyfish in female images. Hydra was a giant swamp monster with nine heads, eight of which were mortal and the ninth, deathless. She had a bad habit of crawling up out of the swamp, like a villain in a grade-B horror movie, to tear apart cattle and lay waste to the land. Heracles kills the deathless head by shooting fiery arrows at it and lopping it off with his sword. Then he pounces on her corpse and dips his arrows into her toxic blood, thus producing history's first chemical weapon. In fact, "toxic" is derived from the Greek word "toxon," meaning arrow. The modern *Hydra*, named after this monster, is a polyp the height

---

* The genus name, *Venus*, cannot be changed on a whim. Undoubtedly *Mercenaria* had "chronological precedence," that is, previous usage.

of the letter *I.* It has six tentacles and is as dangerous to fresh water micro-crustaceans as its namesake was to cattle.

❧❧❧❧ Godly Gastropods

It is not clear to me why snails, in perverse contrast to the phallic worms, inspired masculine names. The common shark-eye snail is named *Polynices,* after the would-be king of ancient Thebes. One of two sons of Oedipus, Polynices used a magic veil to kill mortal women. To see *Polynices* attack a smaller snail is to be reminded of the veil, as the foot of this carnivorous snail expands enormously to form a veil-like pancake of flesh around the shell of the victim, capturing it. The snail then drills a neatly beveled hole into which it inserts its file-like radula, like a giant toothy tongue, to scrape out the contents of the shell until it is empty.

One of the myths about Polynices tells of his victim, Semele, the beautiful mortal princess of Thebes. So lovely was she that Zeus, king of gods, became infatuated by her beauty. He assumed human form to seduce her. The jealous queen goddess, Hera, pursuaded the pregnant Semele to request to see her child's father in his godly form. She was consumed by fire when he revealed himself in his full radiant glory. All of this mythic fooling around has been immortalized. *Semele* is the name of a lovely little clam.

Melampus was the first mortal endowed with prophetic powers. He cared for orphaned snakes, and in gratitude they licked his ear. Upon awakening from a nap, he found that he could understand the language of birds and creeping things. Captured and placed in prison, he heard the "wood worms" talking and told his captors that the building was about to collapse. In gratitude for saving them, he was freed. Though a mortal, he is immortalized as the salt marsh snail, *Melampus bidentatus.* This air-breathing gastropod lives in uncountable thousands on the mud, climbing grass stalks as the tide rises, lest it drown. Several species of ducks survive on a diet composed primarily of this snail. Why Melampus, the mythic magic man, reminded its name-giver of a snail defies my imagination.

PLATE 27

A. APHRODITE, THE GODDESS. The Roman incarnation of Aphrodite, Venus, is shown moments after she was born from the sea. Some experts say that she is disembarking from her shell on the island associated with her, Cyprus.*

B. APHRODITE, THE WORM. The sea mouse, *Aphrodita hastata*, is a segmented worm (annelid) covered with a fine, hair-like coat obscuring 15 pairs of scales. Silky, gold, or bronze, greenish sheen. To 9 inches. May look like pubic region?

C. QUAHOG CLAM, *Venus mercenaria*, named after the Roman version of Aphrodite (now called *Mercenaria mercenaria* because of its commercial importance). The heart shape on the front margin of the shell signifies romance.

D. APHRODITE'S SON, PRIAPUS. This is a Greek Archaic Period fertility idol. One normally thinks of fertility idols as female, but this one is designed to cure what ails males.

E. PRIAPULUS THE WORM, *Priapulus caudatus*, has a thick body of two parts: a white prosoma with a spiny mouth and a thick, warty brown trunk. To 3 inches. The prosoma extends to attack prey—worms and small invertebrates. It becomes turgid when touched. It lives in burrows in sediment. The phylum has only six species.

F. ORNATE WORM, *Amphitrite ornata*, has numerous yellowish tentacles and three pairs of red gills. Reddish-brown body in two parts. Amphitrite was a sea goddess.

G. SALT MARSH SNAIL, *Melampus bidentatus*, is olive with a smooth, translucent shell. The aperture is as long as the body. Tiny, to ½ inch. Melampus was a mortal able to speak to animals. These marsh snails are the primary food of many species of ducks.

H. JOHN DORY, *Zeus faber*, is brown with a conspicuous black spot. To 16 inches. The dorsal fin has eight large spines ending in filaments. Perhaps the crown-like dorsal fin caused it to be named after the king of the gods?

I. MEASLED COWRIE, *Cypraea zebra*, has a glossy chocolate-brown shell covered with white bulls-eye spots. To 3½ inches. It is used as a fertility symbol in primitive cultures because of its resemblance to a vagina. Cypraea refers to Cyprus, the island home of Aphrodite.

* *The Birth of Venus*, Botticelli Sandro, 1485.

PLATE 27

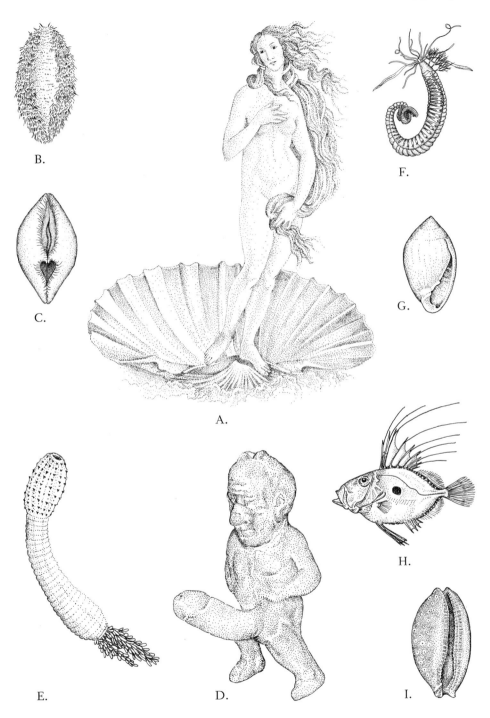

B.

C.

A.

F.

G.

E.

D.

H.

I.

Glaucus was a sea god and seer whose predictions allowed Jason to continue his quest for the Golden Fleece on his ship, the *Argo*. Glaucus and *Argo* are both immortalized—*Glaucus* is an unusual pelagic shell-less snail (nudibranch) that swims through the seas, never sinking to the bottom like its less adventurous brethren. From its sides project fleshy finger-like flippers with which it paddles after jellyfish, several often surrounding one like a pack of wolves. (See page 113.) *Argo* is a squid-like animal called the paper nautilus. It produces a chitinous (plastic-like) "boat" in which it places its eggs, which then "sail" off.

There is one snail that is named after a female goddess—with its resemblance to the female reproductive organ how could it be otherwise? Aphrodite is further immortalized as the cowrie, *Cypraea*, an elongate snail with a slit down its length, resembling to some a vagina. *Cypraea* refers to Cyprus, the island associated with Aphrodite. Love and licentiousness naturally lead to pregnancy, so Aphrodite is also associated with fertility.

The resemblance to the vagina was not lost on cavemen. Women were buried with their most precious possessions, cowrie necklaces and amulets that were thought to assist conception. The presence of this valuable seashell in graves is used to demonstrate that trade occurred among primitive populations. Cowrie amulets are found in graves hundreds of miles inland. Their abundance inland suggests to some archaeologists that commerce transpired between inland and coastal clans of these people. Modern women also favor the vagina-like cowrie shells for necklaces and bracelets. Little do they know of the potential results!

❧❧❧ Fabled Fishes

The big boss himself, Zeus, gives godliness to an olive-green, weak-swimming bottom fish. It is crowned with a conspicuous dorsal fin with spines that resemble a corona. Its scientific name is *Zeus faber* and its common name is John Dory. This name is purported to be a distortion of John Doe, an insulting designation by some fisherman who didn't know his mythology or taxonomy. It is likely that its corona-like crown reminded the taxonomist of Zeus.

Pegasus, that noble god-bearing steed, finds immortality in the name of an ugly fish with a flat, armored body and wide pectoral fins, the sea moth, *Pegasus*. The mythological winged horse soared with godlike heroes on his back to slay monstrous beasts. His birth was a horrible trauma: he burst from the bleeding body of his evil snake-headed mother, Medusa, when she was slain by Perseus.

Pegasus flew on the wind and was a free spirit. Bellarophon, a son of the ever-potent and ever-recurring Glaucus, tamed him and was carried on his back on the quest to slay the monster Chimaera, whose dreadful rampages were bringing death and destruction on the land of Lycia. The loser, Chimaera, gave his name to various species of the boneless ratfish, for example *Chimaera cubana*.

28

Random Ramblings on Relationships

THE SHADOWS of dusk enveloped our boat hidden among the spidery aerial roots of the vast mangrove swamp. Across the muddy, tea-tinged water of the lake the sun still shone brightly, creating a halo around a huge, spreading mangrove tree, its leaves shimmering in the waning light. As we watched, the green slowly became suffused with white. Whiteness begat whiteness as the tree, appearing to be a gigantic snowball, began to melt whiteness into its neighbor. Finally, a third neighbor became infected with spreading whiteness. The ghostly color became amorphous and restlessly flowed from tree to tree in an ever-changing pattern. A cacaphony of hooting and jeering reached our ears from across the lake.

Above our heads in the enveloping dusk, we saw a few white birds, their legs stretched behind them as they laboriously flew by. We recognized them as cattle egrets, *Bubulcus ibis*. The few were replaced by the many; the sky filled with birds. The focus of their flight was the giant white "snowballs." The great white spheres we watched were "egret trees," communal roosting sites for what seemed to be every egret in Trinidad. In an evolutionary sense, it is of survival value to evolve this pattern of choosing mangroves in which to roost and breed. The wastes of the thousands of birds are washed away by the muddy swamp water instead of accumulating in putrid mountains of guano. The surrounding water provides a sanctuary from terrestrial predators.

214

We were awed at the sight, almost hypnotized by the grandeur of the scene. But the show was not over. Act two commenced. A pair of neighboring trees became tinged with red. As we watched, each tree became a gigantic crimson glowing orb in the last of the sun's rays. These fifty-foot-high scintillating spheres were hordes of roosting scarlet ibis, *Eudocimus ruber*, funny-looking bright red birds with thin, downturned conical black beaks used for mucking about in soft, prey-rich sediment.

The method of group roosting in mangrove trees seems to have evolved congruently by egrets and ibis. But the birds are very different. Each dawn draws the egrets toward their scattered stations at the flanks of cattle. They feed predominantly on lizards and grasshoppers stirred up by the movements of their bovine benefactors. Occasionally they return the favor by plucking a fat, bloody tick from a cow. The scarlet ibis depart in small flocks, dispersing among the mangroves and seeking muddy areas that they plumb for small crustaceans.

❧ Conflict in a Monastery

The bird-watching trip ended after dark, and we returned to our "hotel," the Monastery of Mt. St. Benedict, high on a mountain overlooking the Trinidad campus of the University of the West Indies. Dinner was served in the tiny garden.

This was an era of conflict. An unpopular war stirred all Americans. The smell of revolution in the air was not confined to far-away enemies. Internecine conflict smoldered: women demanded equal rights with men—liberation from centuries of male domination. My friends were on the extreme fringes of both movements. Feverish discussion echoed from the brooding ancient walls of the monastery.

One of the young women muttered, "Men are animals," drawing our attention. She raised her voice, drowning out the conversation about the injustice of the war with diatribes against males. "I hate when the pigs stare at me with that hungry look in their eye, as if I were a piece of meat." Then I noticed that her body was enshrouded in many layers of clothing, even on that balmy evening. Her outer garment was a gray sweatshirt several sizes too large, causing her body to appear amorphous. An indistinct furry gray shadow on her upper lip moved up and down with every word she uttered. "My God, is she cultivating a mustache as a symbol of her equality with men?" I wondered.

Naively, I commented that in a biological sense a woman's distinctive organs were meant to attract men for procreative purposes, and that merely by having the requisite anatomy she was fulfilling nature's dictum. What seemed to me a statement of logic almost started a riot among my friends. I looked up at the doorway of the monastery. Engraved on the lintel was the word PAX. I sighed and retreated to my spartan room.*

✎✎✎✎ The Ultimate Female Liberation

Fresh water plankton is dominated by one group of animals, tiny crustaceans called cladocerans. Their prominent feature is a pair of enlarged second antennae that move back and forth, producing an erratic swimming movement. There are four to six underside appendages that act as gills and strainers. These appendages, visible under the microscope, flap back and forth inside a transparent carapace, filtering from the water phytoplankton and bacteria. Many cladoceran species appear to be tiny dots moving erratically through the water, like their brethren the copepods.

The "cutest" cladoceran is the eighth-inch daphnia, *Daphnia magna*. Under magnification a single eye is visible, peering out innocuously from an elephant-like head, the "trunk" being a pointed, recurved anterior projection of the carapace. The heart is visible, pulsating wildly above the dark, food-filled elongate gut that extends the full length of the body. A bubble-like chamber is located just behind the heart. In it can be seen oval spheres—eggs. The hatched eggs release tiny clones of the mother that swim impatiently within the confines of the brood chamber, as if waiting to be released into the fiercely dangerous aquatic world. Few

---

* The Monastery of Mt. St. Benedict is often used as a hotel by visitors to the University.

survive, as daphnia are the primary food of multitudes of juvenile organisms and filter feeders.

Each spring, as the water warms and the bottom layer of the lake awakens, "resting eggs" of the daphnia hatch, and a new generation populates the lake in uncountable numbers. All summer long the daphnia reproduce copiously. Strangely, they produce only females—not a male is to be seen. Summer passes and the all-female population dominates the lake. Inside the brood chamber of the female, cells divide and become two—but they separate into two independent, unattached cells. Each cell goes on to become a new daphnia. Each is identical with the mother; it contains only her genes, not the unique mixture created by sexual interaction with a male. *Males have become superfluous—the ultimate liberation of the female.* No need for useless males.

Fall arrives; the water turns over, bringing nutrients from the bottom to the upper sunlit layers, fertilizing the minute chlorophyll-bearing phytoplankton. The lake turns green. A burst of daphnia population growth accompanies this infusion of microscopic, plant-like single-celled organisms. The daphnia graze on the phytoplankton like cattle in a field, reproducing wildly. The swelling population produces wastes in huge quantities. Toxic ammonia, the cladoceran version of urine, begins to increase, poisoning the phytoplankton and stressing the daphnia. Many die, overcome by the accumulating poison.*

Who will save the endangered all-female population? Miraculously, males appear, as if knights in shining armor. Plankton samples reveal a minor population of male daphnia. Worsening conditions that threaten the population become the stimulus to produce hormones that cause the females to produce male offspring. The males have reproductive cells with only half the normal number of chromosomes. Sperm transfer half-numbered nuclei; they join newly produced half-numbered egg cells. When these sperm and eggs unite, they combine traits from the male and female, producing new variants—the variety that is the stuff of evolution.

Again, only females are produced by this union. But they look different. Each contains one or two huge, hard-shelled eggs. Soon winter arrives. The

---

* In desert-like environments, the cycle is initiated by the advent of the rainy season and the subsequent drying of the temporary pond, instead of autumnal cold.

PLATE 28

A. Inhabitants of temporary ponds and puddles.

Brine shrimps and fairy shrimps, *Artemia salina* (BS) and *Branchinecta paludosa* (FS), swim upside down with 11 pairs of flat legs. Two eyes on stalks. Both species swim slowly and gracefully with coordinated rowing movements. They eat bacteria, phytoplankton, and protozoa by straining them out of the water with setose (hairy) legs and depositing them in a mid-ventral food groove. Both to ⅛ inch. Females have flat egg masses. They usually live close to water's surface.

Copepods (C), *Limnocalanus* (shown) and *Cyclops*, swim in characteristic jerky movements by using their second antennae. Five or six pairs of legs are also used for swimming. From microscopic to 1/16 inch. One red eye, head and thorax fused into a helmet-like cephalothorax. Antennae are important sensory structures and are also used to strain bacteria and protozoa from the water. Females bear egg sacs with five to forty eggs that hatch into nauplii in twelve hours to five days.

Water fleas or daphnia (D), *Daphnia pulex* (see below for details).

Tadpole shrimps (TS), *Triops longicaudata*, glide gracefully with wave-like movements of 35 to 70 pairs of leaf-like legs, but are mostly found burrowing in mud. Two eyes. They feed by using legs to propel detritus into a food groove. Brown or dark green. Huge, may be two inches long.

All inhabitants of temporary ponds produce resting eggs.

B. Water flea or daphnia, *Daphnia pulex*. This species is less than 1/16 inch, *D. magna* reaches ⅛ inch. Carapace is transparent; dark gut runs from mouth to anus. It swims by using a pair of large antennae edged with setae (hairs) that act like oars. The single eye is red. Five or six pairs of setose, leaf-like legs beat in unison to create an internal current that carries bacteria, microalgae, and protozoa into a mid-ventral food groove, thence to the mouth. The legs are not used for propulsion. A dorsal heart, visible near the head, propels transparent blood throughout the body. Behind the heart lies the brood chamber with eggs and swimming juveniles. When ready, they swim out of the brood chamber into the aquatic world, large enough to avoid small predators.

PLATE 28

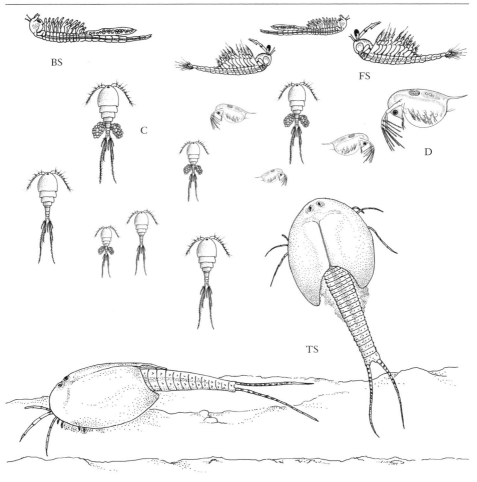

BS

FS

C

D

TS

A.

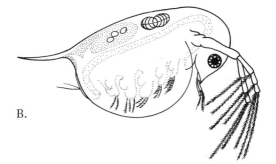

B.

females die and sink to the bottom. The shifting mud buries them, but the "resting eggs" remain, waiting for the warming spring to produce another all-female population.

ᵛᵛᵛᵛ The Rest that Refreshes

The busy bus terminal was crowded with commuters rushing to-and-fro, umbrellas raised. It was thoroughly miserable, one of the many black, rainy days that constitute winter in Israel. The stormy skies were dumping their moisture on us with a vengeance, as if compensating for eight dry months.

A puddle near the bus station was the only ephemeral body of water our city-based team of scientists could think of to search for the inhabitants of what are known as temporary ponds. People stared at the group of oddballs standing in galoshes, pulling plankton nets through two inches of muddy water. Some stopped and commented when they saw us exclaiming excitedly in two languages about the contents of the nets, a crawling, flapping, twitching mass of crustaceans climbing over one another. A few weeks before, this twenty-foot-wide puddle was a dry brown stain on the lawn bordering the main walkway of the bus station. Now it was filled with squirming animals, some an inch long. Where had they come from? "This is the land of miracles," I thought, "but they usually take human form."

A few months or years before, a water bird settled on this spot. It may have landed there only to be frightened away by hordes of commuters (we can identify with the bird) and flown off a few moments after landing. But some of the mud on its feet remained. Buried in it were cysts and resting eggs from the bottom of a drying lake. Thus was life injected into this isolated puddle.

A variety of animals have become adapted to the dry season. In the Middle East an exaggeration of the Mediterranean climate occurs. For eight months it virtually never rains. Land snails produce a coat of mucus over their apertures after climbing up vertical plants, festooning the stalks of dry weeds. They go into a form of arrested metabolism called diapause. All body functions operate at minimal levels, virtually undetectable except with sensitive instruments. But place these "lifeless" animals in a jar with a wet towel so that the humidity rises, and in a few minutes the snails miraculously come to life, crawling rapidly up the glass walls. Instant animals for experiments and teaching.

We were seeking other instant animals, daphnia and other species of the myriad crustaceans that populate temporary ponds and puddles that appear in the rainy season, their populations shrinking, then virtually disappearing during dry months. In the American Southwest, arid areas that have not been wet for years erupt with life after a rare rain.

The most exciting denizens of a temporary pond are *Triops* and *Lepidaurus*, the tadpole shrimps. They are neither tadpoles nor shrimps (demonstrating the need for scientific names). Their inch-long bodies consist of an anterior shield-shaped carapace and an elongate abdomen. On the underside of the abdomen (and thorax) are up to seventy leaf-like appendages that flap back and forth, sweeping in plankton that are carried to the mouth in a ventral food groove. Two diagonal eyes and a pair of modified front appendages embellish the anterior. The elongate abdomen ends in two diagonal spikes (furcae).

An entrepreneur happened upon these bizarre animals and sells them in toy stores as some sort of prehistoric monster. Just add water, wait a few days for the resting eggs to hatch, and presto, your kid has a few *Triops* swimming around in the kit's plastic aquarium. Some species of tadpole shrimps are hermaphrodites and fertilize themselves, never reproducing sexually, an evolutionary conundrum shared with a few races of brine shrimps. Such effective feeders are the tadpole shrimps that they can reach their inch-long maximum size in a couple of weeks. These miniature monsters can appear out of nowhere and disappear as miraculously.

Add another miracle to the traditions of the Holy Land.

29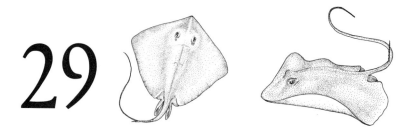

Penile Bloodletting

July, 2004, a marine laboratory on Ambergris Caye, Belize.

The lab director asked, "The class is taking a field trip to Altun Ha (an ancient Mayan Indian city) in a few days. Will you give them your lecture on Mayan culture?" "Sure," I answered.*

After dinner the class assembled, dazed and exhausted after a day in the field. I searched my mind for a "hook" to motivate them. Inspiration! "Today I will DEMONSTRATE penile bloodletting as practiced by the ancient Mayan Indians." Slowly, as the word "penile" was separated from "penal," eyes opened and backs straightened in anticipation. After all, it is not every day that a professor would so mutilate himself on the altar of education.

I began, "The Mayan culture was one of uncertainty and fear. Crops could wither if the corn god, Hun Hunaphu, was not appeased. Floods might accompany the thunderous roaring of the rain god, Chac. The gods must be placated. What can mere mortals do to satisfy them? What is the substance most appealing to the gods, and most crucial to man? *Blood*.

"Ancient Mayan governance rested in a caste of nobility dependent on thousands of stoic peasants to provide them with labor and goods. In return, the Mayan priest-kings were responsible for placating the gods so that natural catastrophes would not occur. The nobility accepted their responsibility by sacrificing their blood, in the most painful possible ways, to the

---

* He knew that I have been fascinated by Mayan archaeology for many years.

222

gods. What is the most suitable implement to show the gods one's appreciation? To the Mayan mind, the threatening, venomous stingray spine. And what is the most appropriate organ to sacrifice in a ritual to placate the gods? To the Mayans—the penis.

"By this logic the nobility paid for their privileges with drops of blood. Ancient murals and carvings reveal their techniques. One picture shows the king passing a stingray spine through his penis. His blood dripped on a piece of paper. The blood-soaked paper was later burned. The smoke floated up to the heavens in supplication on behalf of the Mayan masses. But the queen also had responsibilities. Another scene shows the queen, who, lacking the king's organic apparatus, had to use her tongue. In a famous bas relief she is shown passing a rope of thorns through her tongue."

"TODAY, I WILL DEMONSTRATE THE TECHNIQUE OF PENILE BLOODLETTING." Everyone squirmed. "But, unfortunately, I have demonstrated this one time too often and I physically can't do it anymore. Do I have any volunteers?"

Heads turned toward the men in the class. No volunteers. Then a male latecomer slinked into the class. "Never mind, we have the volunteer." He found out later why his entry into the room was greeted by such raucous laughter. Looking at him fixedly, I fingered the stingray spine. It was exquisite. About six inches long and three-eighths-inch wide at its base, it looked like it was made of ivory. Minute, regular serrations decorated its edges. Bereft of its venomous skin, the minutely barbed spine was still formidable.

When the lecture was over, the class looked deflated. What they had looked forward to with great anticipation had not occurred.

The next day they would have their fill of stingray spines.

᭡᭡᭡ Shark/Stingray Alley

The huge catamaran, aptly named *Goliath*, was crowded with lotion-covered undergraduates. Seaward, the surf created a turbulent white line delineating the protective coral reef guarding the shores of Ambergris Caye. Landward, half-mile away, lay palm-fringed shores. The captain threw smelly fish scraps overboard. Almost immediately, telltale fins sliced through the smooth sea surface, converging on the boat. You could almost hear the music . . . dum-dum, dum-dum. Massive turbulence.

Then, with a ferocious splash, about ten six-foot sharks rose to the surface in a brown writhing carpet. They pressed together, sandpaper sides rubbing in gritty unison. Snouts thrust above the surface, mouths open in grotesque smiling black ovals; they were begging for food like a kennel full of puppies. The captain threw in more food. Loud slurps greeted each handful. Occasionally a ten-pound horse-eye jack, *Caranx latus*, leaped over the feeding frenzy and caught a scrap, virtually tearing it away from the gaping shark mouths. The professor intoned in his best pedantic voice, "These are nurse sharks, *Gingylostoma cirratum*. They have tiny, inconsequential teeth and are bottom feeders, subsisting primarily on sleeping wrasses they slurp out of the sand at night. Now jump in and snorkel to the bottom. I have a surprise for you."

Everyone looked at the professor as if he was crazy. "I'm not jumping in there with those sharks," muttered a bikini-clad young lady. "Follow me," shouted the prof, as he slithered over the stern and snorkeled toward the churning brown mass. I followed, with a string of cautious students behind. Looking upward from underneath, I saw the surface churned to a froth by a mob scene of hungry, jostling sharks. Under the sharks, in mid-water, was a silvery school of about fifty huge horse-eye jacks streaking toward each scrap that fell through the crowd of sharks. As each ten-pound fish whizzed by, it seemed to glare at me with fixed, lidless eyes. Grazing on stringy green algae festooning the boat bottom were hundreds of blue tangs, *Acanthurus coeruleus*, their iridescent blue accents gleaming from the dark shadow cast by the hull.

Captivated by the scene, I sank slowly to the bottom. Movement caught my eye. Gliding over the bottom on rippling "wings," a dozen three- and four-foot-diameter southern stingrays, *Dasyatis americana*, circled in a halo of huge gray disks. Formidable spines angled backward from elongate whiptails. Remembering my dramatic hoax the day before, I could not help but wince. Covered with black venom-soaked skin, the spines clearly demonstrated their appeal to masochistic Mayan monarchs.

The stingrays were attracted by the remainders of the scraps that were raining down like manna in the desert. The professor, kneeling on the bottom, released chunks of conch into the water. He was instantly surrounded by a jostling crowd of stingrays. He waved a punctured Coke bottle filled with essence of conch. It drove the stingrays wild. He swung the bottle in a wide arc. The stingrays followed, creating a moving circle of fish. I couldn't resist. I had to squeeze the "nose" of one. It felt smooth and rubbery.

One stingray followed the bottle closely, almost touching it. But the stingray mouth is located below its flat body about three inches behind its pointed front. It had trouble feeding on the conch bits dribbling out of the bottle. Frequently it would pause to lift its snout and suck up a scrap. These fish are normally skittish, sluggish bottom feeders (as suggested by the ventral mouth), but you would not know it from the eager off-bottom, unnatural behavior caused by this artificial feeding process. Once, off Grand Cayman, in a similar situation, I was given a "hickey" by a stingray that confused my elbow with food. She must have been hungry because that kiss was as passionate as any I have received.

ᨒᨒᨒ Death and the Stingray

The southern stingray, *Dasyatis americana*, annually stings a surprisingly large number of individuals, mostly in the Gulf states where it is common to stalk the shallows at night, spearing flounders. The stingray is typically buried in the sand. When stepped on, the tail reflexively swings upward, the spine extending almost vertically. The venomous skin enters the shin or calf and tears, exposing the serrations along the spine edges, making removal difficult. In the dark of night the attack is frightening. The first symptom is intense pain. The pain becomes more and more intense for about thirty minutes.

An ambulance brings the victim to the hospital where procedures to prevents shock are instituted. Antivenom is not used and the symptoms are dealt with as they appear. Swelling is followed by tissue damage (necrosis), especially when the spine is torn out of the wound. Blood pressure drops. There can be nausea, faintness, dizziness, and shock symptoms. Children are more vulnerable than adults. Among many deaths reported, one tragedy is particularly sad: an eleven-year-old boy in Texas died two and a half days after being stung on July 4, 1954.*

In the field, cold water or ice should be applied to the lesion to slow down the blood flow, and a tourniquet started. The wound should be examined and any remnants of the toxic skin torn from the spine should be removed.

The six-inch stingray spine resembles an elongate pointed tooth with an inner dentine and outer enamel layer—a miniature elephant's tusk. Venom is produced in glandular tissue buried in a groove sculpted into the enamel near the edges of the spine; hence its name, ventrolateral gland. When the sting is embedded, its skin, containing the gland, tears off and remains in the wound.

Fortunately, the Mayan Indians knew enough to remove the skin from the spine before applying it to the penis.

〜〜〜 Jaws

Perhaps the only person ever harmed by a nurse shark was the "world's largest and funniest marine biologist" (see chapter 2). Here he was, standing in five feet of water, holding a six-foot nurse shark up for his students (in the boat) to kiss. So large is he that the nerve impulses coming from his legs seem to dissipate before reaching nerve centers in the brain, so he doesn't feel pain in his extremities. A female student indicated that she had an urge to kiss the shark, but her pouted lips did not extend enough to reach the fish as she recoiled with fear. World's largest prof had compassion for the young lady and held the writhing shark up ever closer. The young lady finally screwed up her courage and popped a smooch on the snout of the shark, giggling mightily. It was only on the way back that he noticed that his legs were red with blood. The shark's sandpaper skin had rasped away his skin as he played the kissing game.

* Bruce Halstead, *Poisonous and Venomous Marine Animals of the World*, (Ephrata, PA: Science Press, 1978), 653.

Nurse sharks are so sluggish that when they lay about in groups during the day, scattered over the bottom and piled up on one another, they appear as inert as logs. Unlike many sharks that must swim constantly to flush oxygenated water past their gills, nurse sharks use a muscular canal with an opening behind each eye, the spiracle, to push water over the gills. So sluggish are they that this method is more than adequate to provide them with oxygen.

They feed at night and rest during the day, exhausted from sucking wrasses from the sand with a bellows-like pharynx. (They eat other bottom dwellers like crabs and octopods—and stingrays.)

❧❧❧ When Nurse Sharks Become Lively

All is still among the scattered nurse sharks. Occasionally one stirs, but only to relieve the pressure from the weight of one of his comrades sleeping, wide-eyed, on top of him. Then suddenly, there is a mass stirring. A receptive female cruises by, her pheromone perfume assailing male olfactory receptors that can pick up scents from miles away. She moves slowly over the bottom. The males, in their sluggish way, are in a sexual frenzy. Several approach the female. One moves alongside her and, flank to flank they swim off. In a moment of slow-motion sexual abandonment, he grasps her pectoral fin in his jaws (scarring her fin with tiny tooth marks) and tries to turn her sideways. If she is not receptive, she will swim into shallow water and bury her pectoral fins in the sand, rebuffing him.

But eventually she succumbs and he inserts one of his claspers into her cloaca. He uses the other to clasp her to keep her close. Each clasper is actually a modified pelvic fin, with a groove to carry the seminal fluid and a spiny tip to embed in the female's uterus to insure insemination. Sperm run down the clasper and into her oviducts, where fertilization occurs. A few months later twenty to thirty olive-colored, flask-shaped, three-inch egg cases line her uterus. The pups hatch inside the uterus, each attached to a yolk mass. The female provides no other source of sustenance in this classical case of ovoviviparity and the pups are born nine months later, able to feed.

PLATE 29

A. Stingray spine without poisonous skin. Ivory, to 7 inches long by ⅜ inch wide.

B. Scene under the boat, Shark/Stingray Alley, Belize. The boat captain throws chunks of conch into the water and the surface is churned to a froth. Nurse sharks crowd together, their pointed snouts breaking the surface. Beneath the sharks is a school of horse-eye jacks. Beneath the jacks, the sandy bottom is paved with stingrays. The jacks catch most of the chunks that escape the sharks and the stingrays are left with sparse remainders.

Nurse shark, *Gingylostoma cirratum*, in a feeding frenzy at the surface. Brown or gray with a pair of barbels adjacent to the mouth near the nostrils, slanted yellow eyes. Two large dorsal fins, the second smaller than the first. To 12 feet.

Horse-eye jack, *Caranx latus*. Tail yellow, body silvery, bluish above. Huge eyes, blunt head, row of black ridges near the tail; 12 to 24 inches long.

Southern stingray, *Dasyatis americana*. Brown or gray above, white below. To 5 feet wide. The mouth is on the underside, well back from the pointed snout. It swims by undulating its fins. The tail is long, narrow, and flexible with a 6-inch spine.

(Below) Bas relief showing King Pacal, ruler of Palenque handing ceremonial stingray spine to his favorite son, signifying son's succession to throne.

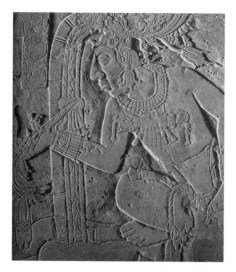

PLATE 29

A.

B.

✢✢✢ Another Kind of Nurse Shark

The unrelated Australian gray nurse shark, *Carcharius taurus*, feeds on small fish with a mouth that can be thrown out five inches to engulf prey. It was systematically slaughtered as a "man-eater" until it was discovered that nothing larger than the small mouth can be engulfed. Its unique style of reproduction makes it famous. Fertilized eggs are deposited in each of two uteri. Inside a uterus, an egg case disgorges a fetus. After further development, it becomes active inside the uterus *and eats the other developing embryos*. This intrauterine cannibalism continues until all the embryos are devoured. Then the mother, in unwitting encouragement of this cannibalism of her young, produces unfertilized, yolk-rich eggs that are promptly eaten by the juvenile, sustaining it until birth. Eventually, the other uterus produces a mature juvenile in the same manner. Mother gives birth to twins! Although mom provides sustenance in this case, it is not considered an example of viviparity because, in the narrowest sense, a placenta is required to achieve true viviparity. Remarkably, sharks exhibit all three forms of child care: some lay eggs in cases entwined by tendrils on rocks or corals; some keep the eggs inside the uterus, offering no sustenance; some use a placenta.

✢✢✢ No Sting Needed

The boat rounds a bend to reveal a cove right out of a travel magazine. (As a matter of fact, this cove is about a ten minute boat ride from the beach made famous by the movie *Doctor No*.) We stared at the gorgeous palm-fringed Jamaican shore, half expecting James Bond to appear and hoping that a beautiful woman would wash up in the quiet surf. Alas, we were disappointed. The class scattered, the less-skillful snorkelers floating face down in about three feet of tranquil water, following tiny jewel-like fishes and gold-spotted crabs. The macho men and women thrust toward the sea bursting in white spume against a black rocky promontory protecting the mouth of the cove.

I worry about the exuberance of youth and its efforts to prove something by placing oneself in danger. This beautiful cove was tranquil and safe. Why did they have to go beyond the boundaries? We followed, ready to haul in a waterlogged undergraduate. Suddenly my ubiquitous graduate student pointed to the bottom. "Is that *Urobatis* or *Narcine*?" he asked. Both small rays are rounded, both are tan with brownish spots. "Stay here, I'll let you know." I dove toward the ray, my hand extended. When I was about a foot away, I felt a distinct shock. My arm tingled. "It's *Narcine*," I spouted upon surfacing. The tingling subsided after a second or two.

Narcine brasiliensis, the lesser electric ray, paralyzes small fishes by creating a mild electric field of 14–37 volts, hardly appreciable by man. It needs no sting, as it warns off predators with an invisible electric aura. The electricity is created in a pair of electric organs comprised of five hundred flat, plate-like modified gill muscles in a honeycomb pattern. The same process used for movement is subverted to generate an electrical field. Some electric rays and catfishes can produce currents of 500 volts, enough to shock a human.

A pair of gray, bumpy, kidney-shaped electricity-producing organs are clearly visible behind and to the sides of the eyes of *Narcine*, differentiating it from *Urobatis*, the yellow stingray.

The ancient Greeks called *Narcine* the "numbfish" and used the shock to induce numbness or electroanalgesia. The Greek word for numb, *narke*, is recognizeable in narcotic and *Narcine*.

꥟꥟꥟ Bad Habits

We all go through a period of being BLBs (or BLGs) (see chapter 21). My particularly nasty version, at age twelve or thirteen, involved catching a skate (similar to a ray, but without the sting), turning it over to expose its white, vulnerable underside, and putting a lit cigarette between the soft, human-like lips. While gasping for water, the poor thing would appear to be smoking. I feel bad now, because it could have died of gill cancer after we threw it back.

30

Death and Confusion

ROLL CAMERA. THE handsome, muscular star walks into the shallows. Cameras follow every move. A microphone on a boom records every sound.

Ominous shadows are visible everywhere in the shallow waters. This is Bull Shark Cove on one of the Bahama out islands, Walker's Caye.

A spotter is stationed behind the star. His sole responsibility is to distract an overly aggressive animal by throwing a dead fish far away from the star.

A huge female bull shark moves too close. The spotter yells and throws a fish chunk. Unfortunately, a remora detaches itself from the shark and bites the chunk with a mouth too small to swallow the smelly distraction, bumping it and pushing it close to the star as he stands hip deep in the shark-roiled muddy water, explaining how sharks will not attack humans unless threatened. The huge shark attacks! In a frenzy of thrashing, man and beast become as one. Blood tinges the water red. The shark darts off, leaving our TV star frozen in shock, bleeding from a wound that shears off his calf muscle. The camera rolls on.

This is not a replay of *Jaws*. It is a real made-for-TV documentary gone wrong. The "star" is, in reality, a controversial scientist, an expert in shark behavior. Instead of sitting in a lab theorizing, he was testing his hypotheses in God's laboratory, the open sea.

232

❧❧❧❧ Hysteria

The injured scientist was medivacked to Miami. He was saved, but no matter how much surgery he was to have, there was no way to replace the functions of his calf muscle. His skiing days are over, his Swiss heritage forsaken.

Something extraordinary happened after the accident. A flurry of articles appeared in local newspapers and scuba magazines. They were hysterical attacks on the unfortunate victim, with titles like "Evil Pundits of Doom: Idiot of the Week."*

The TV program aired on major channels, provoking even more angry hysteria. What can we make of this massive, violent, negative response? Instead of compassion, people savagely attacked the injured scientist in a feeding frenzy of media rage.

What is the explanation for this near-maniacal public outrage?

The original purpose of the TV program was to demonstrate that shark behavior is predictable and that if one is familiar with sharks' responses to stimuli, one could move among them with impunity. This was a hypothesis based on years of underwater experience and observation. The shark biologist had spent many hours underwater with shark schools and individual sharks. In fact, there is a body of literature categorizing their response patterns to threats, invasion of territory, and other stimuli.

When this message "went bad," amateur shark aficionados and scuba enthusiasts saw one of their pet beliefs about sharks undone. They *wanted* sharks to be harmless, unthinking animals. This bloody demonstration had revealed a contradiction in their cherished beliefs—like disproving that apple pie stands for America.

For centuries sharks have been feared as predators of people. As a result, their populations were decimated. Nowadays, rapacious human predators

---

* The reader should know that the star of this true story is a friend of mine. Is this an attempt to vindicate him, or am I trying to explain how the public reacts angrily to any threat to its cherished beliefs?

of sharks, made brave by new technology, systematically slaughter them. Not out of fear do shark hunters kill millions of sharks each year, but for soup.*

The shark protection community of amateur naturalists and scientists knows that sharks suffer from the same weakness as all top predators— they have low biotic potential (produce few young). The overly emotional response comes from the knowledge that if the slaughter continues, some sharks will be in danger of becoming extinct. This one shark bite, smeared over prime time TV, was living proof that sharks are dangerous. Shark biologists, naturalists and other interested parties† worried that this one incident would fuel the fires of fear and cause even more shark destruction. That seems to be the reason that a hypothesis proven invalid provoked such negative hysteria. Shark behavior ranks with stem cell research as an issue in which science confronts fanaticism in segments of the lay public.

If we treat this incident with the dignity it deserves, we can examine other hypotheses to explain the shark's unexpected behavior. First, we must understand one maverick shark cannot undermine the hard-won idea that sharks respond in a stereotyped manner that can be predicted. This shark's unexpected response should be judged against years of observation on hundreds of these fishes that did behave in the expected manner.

There are alternate explanations for the aggressive actions of the shark. Was it destabilized by the movement of the many human legs in the water at the same time, those of the team of technicians involved in the production? Or did the cloudy water obscure its vision? Or did the presence of the bloody dead fish chum cause it to strike at anything in its path?

But the most likely explanation is that the shark was confused.

---

* Each year millions of sharks are captured for the purpose of obtaining fins for "shark fin soup," an Asian delicacy. Fins are cut off and the remains of the dead or dying sharks are thrown overboard.

† "Now that shark diving has become a multimillion-dollar business, a living reef shark can be worth a fortune. According to one study, a single reef shark is worth $100,000 every year to local businesses." Mark Cowardine and Ken Watterson, *The Shark Watcher's Handbook* (Princeton, NJ: Princeton University Press, 2002), 17.

❧❧❧ Confused Sharks

I have known several so-called shark experts who dive into schools of sharks in the belief that they are immune to shark attacks because of their knowledge of shark behavior. One of them actually traveled to South Africa where there is an area notorious for its great white sharks. Forsaking the traditional shark cage, he dove right into a sea previously seeded with bloody fishes to attract the sharks. He was following the philosophy of his guru, an animal behavior expert who did not test his hypotheses in person, but used the gullible scuba enthusiast as a guinea pig to find out if his ideas about shark behavior were true. I have not heard from him in many years. I wish him well.

In contrast to the guru, our hero had the guts to put his life on the line to test his hypothesis that sharks do not attack humans as prey. This hypothesis was based on years of observing shark behavior. He suggested that shark attacks are aberrant phenomena.

Was his hypothesis wrong? Was this unfortunate event a rejection of years of observation? Circumstances surrounding the event need to be examined.

The reason that the TV program was being filmed off that island, in that cove, is that a scuba resort had used a simple technique to bring the sharks there and keep them in the vicinity. Every day the resort owners hung a barrel of frozen fish on a line. When the ice melted, the bait was freed from its icy prison and drifted downstream to the waiting sharks. They left the dark depths and drew closer and closer to the source of the heady smell dangling in very shallow water. The sharks had learned that a free meal was to be found in Bull Shark Cove. A "free meal" to predators is one that requires the expenditure of as little energy as possible. Lions rarely attack healthy animals; they are more successful with injured prey or weak juveniles. Wolf packs surround a moose trapped in the snow. All fishermen know that a lure that pops along the surface will call in predatory fishes because it imitates wounded prey, easy pickings.

A logical hypothesis is that the six-foot bull shark, *Carcharhinus leucas*, lunged for the food and missed, accidentally severing the adjacent human calf muscle in its confusion. Bull sharks attack humans all over the world. There is a fresh-water population in Lake Nicaragua that migrates up the

San Juan River to and from the Gulf of Mexico. Every year people are bitten by these sharks.

Bull sharks are rotund. A bulky six-footer may weight 350 pounds. But its mouth is relatively small. It has to tear its prey apart into pieces to devour it. Yet in virtually all cases of bull shark attacks, the human victim is bitten only once, as if human flesh doesn't taste good to the shark. In this instance the telltale fact that the shark fled after its only bite suggests the simplest answer, that *confusion*—not predatory intent—explains the attack. How is it that bull sharks so consistently bite humans? Fishes respond to prey size. Turtles, alligators, and other sharks are often found in the stomachs of these sharks. That is, they normally attack large prey—and in rivers and lakes, where most bites occur, the water is muddy. Potential victims are blurry. Mistakes result in ripped calf muscles or severed legs.

Great white sharks, *Carcharodon carcharias*, accidentally kill people. Their most important food is seals and sea lions. From underwater, humans bear an uncanny resemblance to these animals. All three produce slow swimming movements distinct from those produced by fishes. The sharks home in on their characteristic sound patterns. Sea lions spend much of their time swimming off beaches. Humans swim there, too. If a thousand-pound great white makes a mistake, the human is dead, even if the shark spits him out in revulsion—which it does not seem to do that often. Great whites are not picky. Tin cans, buoys and dead people have been found in their stomachs.

◈◈◈◈ The Inner Shark

Sharks are primitive animals, having appeared hundreds of millions of years before the ancestral fish line split into two forks: the elasmobranchs, the cartilaginous fishes; and the teleosts, the bony fishes. Sharks dominated ancient seas three hundred million years ago, long before bony fishes appeared.

Sharks, skates, rays, and remotely related chimaerids (ratfishes) have no bones. Support is provided by cartilage. It is more dense and solid than the cartilage at the tip of your nose and in your ears. Sharks lack swim bladders used by bony fishes—balloon-like flotation structures containing air secreted by a special organ. Although sharks use copious amounts of oil for floatation (oil floats on water), most must swim constantly to cause oxygenated water to flow over their gills. Nurse sharks use a system of muscular structures and valves to provide life-giving oxygen, but their metabolic needs are not great. They spend their lives in sluggish repose, feeding at night on vulnerable, easily obtainable prey.

❧❧❧ Sharks Have Brains, Too

The three parts of all brains, the hindbrain (medulla), midbrain (cerebellum), and forebrain (cerebrum), are enlarged according to their primary function. In humans the modest medulla controls the autonomic functions such as breathing and intestinal motility. In the shark, it is swollen with other functions. It receives sensory input from an acoustical system, passing it on to the auditory lobe. The shark "hears" vibrations in the water with a system of sensory nerve endings organized into a row along its flank, called the lateral line system. Another unusual input into the medulla comes from a group of electroreceptors in the shark's head. Thus, in proximity to prey, it can detect the victim's electrical field. Anterior to the medulla is the cerebellum, the center of balance and coordination in the human. It has this function in the shark, too, receiving input from the lateral line system.

The shark's forebrain differs from its human counterpart in its disproportionately swollen optic and olfactory lobes and reduced size and function of its cerebrum. The frontal lobe of the cerebrum, the huge thinking part of the human brain, is virtually absent. But recent research suggests that a substantial part of the shark forebrain is given over to mysterious cells that may function in memory and learning. Absent a clear explanation of these cells, we will follow the traditional belief that the brain of a shark is primarily given over to sensing vibrations in the water, electroreception, and, to an overwhelming degree, smelling. Historically, sharks have been described as "living noses." It is true that sharks can detect the scent of blood from a mile away, but it is a better measure of olfactory sensitivity

PLATE 30

A. Bull shark, *Carcharhinus leucas*, competing with a remora (shark-sucker), *Echineis naucrates*, for a chunk of fish. To 11 feet. Silvery or dark gray with brown dorsal surface. Heavy-bodied with a short, broadly rounded snout and tiny eyes. Has a large triangular, pointed first dorsal fin with a very wide base.

The remora reaches 3 to 4 feet long. Dark brown on top with a broad, blackish stripe outlined in white on its side and a large corrugated sucker on its head. Remoras are found on a variety of fishes besides sharks, including marlins, barracudas, and jacks. A remora hitchhikes a ride on the host, and detaches and feeds when chunks of the host's prey float by. Sometimes it will not relinquish its hold on the host even when it is hauled aboard a boat.

In this scene, the remora bumps an oversize chunk toward the scientist, and the shark bites the scientist by mistake.

B. Threat display postures.* This behavior is exhibited by virtually all members of this species, the gray reef shark (*Carcharhinus menisorrah*), under a variety of threat situations. The display postures are on the left, the normal appearance on the right. Under threat, the body arches and twists—a string of automatic, unthinking reflex-like behaviors.

C. The brain (white) of the spiny dogfish shark, *Squalus acanthius*, compared to a human brain. The human cerebrum dwarfs that of sharks, but the olfactory and optic lobes are proportionately larger in the shark. The swellings from left to right: olfactory bulb, cerebrum, optic lobe, cerebellum. The shark has sacrificed thinking capacity for highly developed senses of smell and vision.

D. Shark's eye view of a swimmer on a surfboard. Great white sharks, *Carcharodon carcharias*, feed on sea lions, seals, and large fishes. The silhouette of a surfer on his board may resemble a sea turtle or the mother of all sea lions to a cruising great white shark. The species feeds on active prey, so it has done away with the sensory approach sequence and rushes to attack lest its prey escape.

---

* R. H. Johnson and D. R. Nelson, "Agonistic Display in the Gray Reef Shark, *Carcharhinus menisorrah*, and its Relationship to Attacks on Man," *Copeia* 1 (1973): 76.

PLATE 30

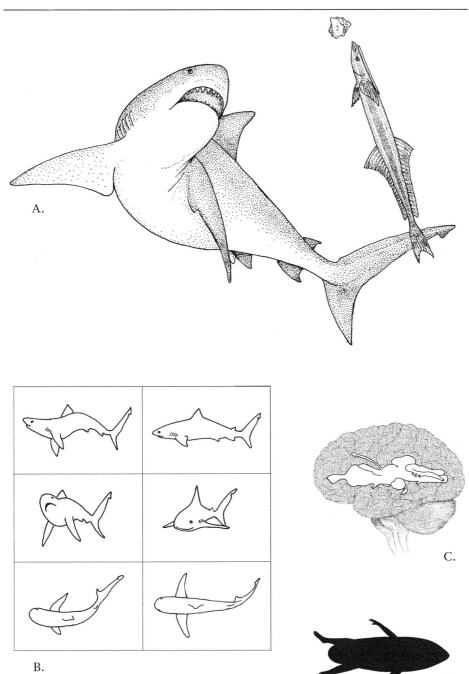

A.

B.

C.

D.

to explain that the shark's ability to detect an odorant is one thousand to ten thousand times ours. The animal thus becomes the ultimate aquatic predator.

﹌﹌﹌ Do Sharks Think?

Some shark species assume a specific posture when a certain size object enters their "personal space." This posture is termed an "attack" mode. The body assumes a humped appearance.

Can this complex behavior be an example of a string of innate, automatic behaviors (taxes)? Physical and environmental conditions can make the response seem variable. Distance of the object from the shark, size of this potentially dangerous "foe," and clarity of water affect the response, making it appear that the shark is thinking. But if these factors are ruled out, is the response invariably the same or is it really variable (thinking)? Our hero was testing the hypothesis that the behavior is invariable.

In the inner recesses of his mind, he may have remembered another aquatic predator, the leech, a predaceous relative of the earthworm. A leech smells the presence of a human in the water and swims toward her. Then it detects its prey with four anterior light-sensitive spots and attacks the dark shadow of the human's calf muscle. The horrified human victim, seeing this, runs screaming out of the water. But the brain of a leech is a ring of nervous tissue running in a never-ending circle around its pharynx. At the top of the circle is a tiny node of nervous tissue that responds to smells and shadows. Is this glorified earthworm, performing these shark-like behaviors, thinking?

﹌﹌﹌ Shark Attack

The hypothesis: The absence of a cerebral cortex suggests that sharks are not heavy thinkers. Like most fishes, they cannot reason, but they can learn to avoid threatening stimuli and to remember where sources of food can be found. Their behavior is governed by taxes—automatic, inborn responses *primarily determined by their genes*.

Our hero suggested that the feeding behavior of most sharks consists of a ritualized series of taxes: First come the olfactory stimuli—the shark homes in on certain smells. These may not necessarily be blood, but body secretions like a stream of urine or body oils.

Then the "auditory" (lateral line) system comes into play. The vibrations created by a school of tuna or the characteristic pulsed, low frequency waves of a seal or turtle bring the shark closer. It sees its prey. Light colors are more obvious than dark. (In World War II yellow or "international orange" life preservers served only to attract sharks. The color was eventually termed "yum-yum yellow.")

Drawn close to its prey, the "near-field" senses are used. A few inches from the potential meal, the shark picks up its electrical field. Then another sense helps it to decide to attack—touch. If possible, the shark will bump the prey with its body. If all stimuli are "go," the shark attacks. This ritual is invariably followed by most sharks, great whites not included. Since they feed on highly maneuverable seals and sea lions, great whites cannot afford to follow the ritual. They respond to visual stimuli, speeding in a direct line towards the prey. Confusion between a sea lion and someone on a surfboard is therefore more likely in this species.

Given our knowledge of the disproportionate attribution of brain cells to the senses and the lack of a cerebral cortex, is shark behavior invariable, a series of automatic responses? *Sense the prey, go through the ritual.* But the bull shark that was the inadvertent subject of the TV show did not undergo the expected series of behaviors. Does this mean that it was "smarter," less reflexive, than its brethren in the area that did not attack? What do you think? I opt for this hypothesis: its unnatural behavior happened because it was confused.

31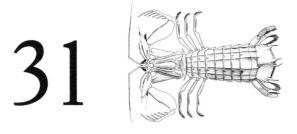

Eyeball to Eyeball

HE SASHAYED INTO THE ROOM, Hawaiian shirt reflecting glorious underwater scenes. A lock of unruly blond hair partly obscured one eye. Extending a muscular, tanned hand, he greeted me warmly. As the class came to order he chose a seat and was immediately surrounded by adoring coeds.

After class he wanted to chat and we went to my office. He explained his unusually informal social behavior. It seems that he had been a navy diver until an accident forced him to seek a landlubber's life. But the sea drew him like a magnet and he washed up on the shores of Hawaii. Using his navy expertise, he established a scuba diving school and prospered. He was happy in his paradise of bikini-clad students and the world of corals and gorgeous tropical fishes. "That sounds like every man's dream. How come you are here in my invertebrate zoology class?" I asked. "I got to thinkin' about how I wanted to know some more about what I saw underwater, and then I got a bug up my butt to become a marine biologist."

I explained to him, as I have to you (see the preface) that there is no such thing as a marine biologist, and anyway, it takes years of study and a PhD to function in the scientific world as he wanted to. I couldn't help saying what was then the truth: "I'll change jobs with you in a second." We both laughed at the incongruity of the idea (but there was a hollow ring to my laugh). "This is a tough class, do you feel prepared for it?" "You belt it out, boss. I'll catch your passes and run with them."

242

Time passed and he proved to be intelligent and able to grasp complex concepts. The trouble was that he was able to parlay his previous experience into comprehending material he had dealt with in his diving career, but unfamiliar topics requiring discipline were not his forte. In short, he didn't study.

It came time for the final exam. I was unhappy for him because I realized that his lack of discipline would cost him his dreams, at least in my class.

Dutifully, he sat in the first row for the two-hour exam. He left after one hour, giving me a wry smile. I looked at his paper and most of the questions were not answered. Later, during the grading process, I sadly went over his exam booklet. In one section, the page was blank, except for one answer. The question was "define ommatidia." The correct answer is "the tiny photosensitive units that compose the compound eye of arthropods." His answer was *"what the lady said when her chest got stuck in the door."* Gasping with laughter, I gave him five points for creativity. Alas, even with this brilliant contribution, he didn't pass. His was the funniest test answer in my long career. It cheered me up during bouts of frustration at the dull exams I graded over the years.

꽃꽃꽃

Ommatidia are the culmination of vision among the arthropods. Animals possessing them can "see" in the sense that visual images are created. But they are an end, not a beginning. Where did it all start? The stepping stones in the evolution of vision exist today, so we don't need a fossil record to follow the laborious, slow increments leading to the kind of eye that can create real images. Some of these optical steps have achieved considerable complexity, but they still cannot "see." No image is created— they remain precursors no matter how universal their distribution among invertebrate animals.

᭣᭣᭣᭣ To See or Not to See

The evolution of vision is marked with monumental behavioral and ecological consequences. Predation became profoundly more effective when sight enhanced the capability of the predator. In fact, the capacity to see may have caused the downfall of radial symmetry.

Protoplasm itself has the capacity to respond to light stimuli. The familiar protozoan, *Paramecium,* responds to light without any light-sensitive structures. Light perception at this one-celled stage begins a stepwise evolutionary sequence. Although they consist of only one cell, some protozoans evolved specific light-sensitive structures. *Euglena,* a flagellated protozoan, uses an eyespot. This microscopic dot-like organelle makes it possible to exhibit phototaxis, an innate response to light.

The tidal range of the Thames River passing through London on its way to the sea is such that it is drained during low tides. An unpleasant, smelly mud bottom is all that is seen when crossing London Bridge.* Then, as the sun reaches the mud, its surface begins to shimmer with a gold-green sheen. A prodigious population of one-celled *Euglena* migrates from under the surface to expose chloroplasts to the sun. Evidently it is of survival value for them to populate the mud invisibly until the sun calls them forth. Then flagella undulate, propelling them to the surface.

᭣᭣᭣᭣ Red-Eyed Dealers of Death

Red eyespots are commonly found in flagellated protozoans. In shallow seas the ocean literally turns red as almost infinite numbers of tiny red "eyes" peer out of the sea when conditions are right for reproduction—often when certain pollutants are present. These red "eyes" are the light-perceiving

---

* The original London Bridge was saved from "falling down" by an American entrepreneur. It is now in Arizona.

structures of the dinoflagellate, *Karenia brevis* (=*Gymnodinium breve*). When the water is thick with inconcievably huge populations of these dinoflagellates, they create the fearsome red tide. Billions of *K. brevis* fill the sea with paralysis-causing neurotoxins. The effect is caused not only by their high concentrations in the water, but also by biological magnification.

Clams normally feed on phytoplankton. They cannot differentiate their food except by particle size. But *K. brevis* is as small as any other phytoplankton and the clams cannot distinguish it from its cohorts. Every day they strain vast quantities of *K. brevis* out of the water, concentrating substantial quantities of its paralytic toxins in their tissues. To eat these toxic raw clams is to gasp for breath as the concentrated toxins impair diaphragm movement. Sometimes hospitalization is required. Similarly, the bay, previously rich with unseen aquatic life, becomes a cemetery for thousands of paralyzed gasping fish that literally cover the water's surface.*

Eyespots cannot perceive images, but sensitivity to light permits awareness of the sun-drenched surface and its promise of photosynthesis. Some dinoflagellates descend as far as two hundred feet at night to eat bacteria, and rise to the surface during the day to carry on photosynthesis.

ᜧᜧᜧᜧ The Next Step

Multicelled animals continue the sequence. Lowest on the evolutionary tree, cnidarians (jellyfish) and ctenophores (comb jellies) are covered with scattered light-sensitive cells associated with the most primitive nervous tissue. Cnidarian and ctenophore nerve cells form a network that does not coalesce into a central control structure. This nerve net is merely a tracery of cells that carry nerve impulses—to where? There is no central coordinating brain. The impulses simply go around and around, stopping who knows where. To be fair, there is some suggestion of aggregation of

---

* Salmon farms are forced to harvest too soon when tests reveal that a red tide is approaching, at a loss of millions of dollars. Similarly, commercial fishing is prohibited during red tides.

nerve cells along the edges of the mouth and a significant aggregation of these cells at the aboral end of the comb jelly. But again vision is a matter of differentiating light from dark, not fundamentally different from the protozoa.

Some jellyfish have taken the next, monumental, step. They have evolved an accumulation of these light sensitive cells called an ocellus. Virtually all ocelli contain dark pigment whose purpose is to shut off all light except perpendicular rays, like primitive Polaroid sunglasses. Vision is impossible; lenses to concentrate light are lacking or, in some cases, present but too primitive to be effective. Ocelli are useful anyway. They can more clearly differentiate light from dark. Free-living flatworms may have rows of light-detecting ocelli along their margins. *Diadema*, the long-spined black sea urchin, peers with its ocelli from between its spines and aims them at the shadow of a diver.

Annelids, segmented worms, have ocelli concentrated into two or four "eyes," but the "eyes" cannot detect images and are used only to determine the darkness of night, when the worms can emerge from under the sand to forage safely. I have seen that sinuous predator, the *Nereis* clamworm, explosively evert its two-fanged fleshy pharynx into the side of another worm as if it could see its prey. But electron microscope examination reveals that its four "eyes" can't focus, and it feeds by feel.

The scallop, *Aequipecten irradians*, whose adductor muscle we eat, peers out from beneath the undulating edges of its shell with dozens of complicated baby-blue eyes, but still can't see. The red-tide organism *Karenia* has a red "eye." The *Nereis* clamworm has black eyes. Blue, red, black—it seems that color is of significance in seeing. Vision begins with nerve cells specialized to respond to light. Their excitation creates nerve impulses that are carried to the brain for interpretation. These cells are almost invariably pigmented. It seems logical to think of color as a means of absorbing maximum light and harnessing that energy into an impulse. The most primitive eye is just an accumulation of black-pigmented cells, spots on the skin. More advanced, some ocelli are "eye cups," darkly pigmented cup-shaped accumulations of photosensory cells. In the next evolutionary step, transparent lens-like structures appear in the center of the cup.

The most advanced eye cups have double layers of black-pigmented cells, like the human retina. Any light that is not absorbed by the innermost layer bounces off and is absorbed by the outermost layer on the way out in this efficient ocellus.

But still no distinct images.

❧❧❧❧ The Miraculous Compound Eye

A profound step beyond the ocellus is the compound eye. It is capable of real vision, but of a nature different from ours. Made up of hundreds of tiny visual units called ommatidia, vision is not concentrated into one image, but is a mosaic of many repetitive images. Each image is vague, but this step up represents the culmination of vision in the arthropods. Survival is at stake. Some claim that Arthropoda is the most successful phylum. The compound eye must make an important contribution to the phylum's success. Why not go to the next step, a refined single image like ours? Why settle for many vague images when a single, precise image beckons? The answer is that each of the massed ommatidia produces its own image. If a predator or prey animal moves, hundreds of images flit from ommatidium to ommatidium. Detection of movement, rather than visual precision, is the ultimate refinement of the arthropod eye. Now it is possible to explain the magic of the fly's ability to detect the swatter from any direction and elude your determined efforts.

❧ The Color Champ

I am searching for brittle stars under coral rubble. Lifting a dripping rock, a head-sized patch of shade is uncovered. A horde of startled skinny-armed brittle stars is exposed, causing them to explode into-activity. Frantically, they sinuously clamber over the bottom, their ocelli suddenly bombarded with light. I slip my hand under the specimen I want, knowing that if I try to pick it up by one arm, it will autotomize (break off) that arm instantly, and I will be left with a writhing string-like appendage.

PLATE 31

A. MANTIS SHRIMP, *Squilla empusa.* To 10 inches, grayish-white, edible, but rarely seen due to subterranean habits. Common in northeast U.S. coastal waters. The attitude of prayer disappears when prey approaches. In a split-second, a front appendage swings upward and slices the hapless prey in half. Other mantis shrimp punch their prey to death. Eyes are black diagonal structures at the base of the antennae. Mantis shrimp (stomatopods) have the most discerning color vision of all animals, and it is difficult to understand why they have such unusually sophisticated eyes, likely the pinnacle of evolution of the compound eye.

B. GHOST CRAB, *Ocypode quadrata,* has upright, stalked compound eyes composed of a few thousand ommatidia, not hundreds of thousands as in the mantis shrimp. Insects, crabs, and shrimps have hexagonal ommatidia. Each ommatidium consists of a cornea/lens and a layer of pigmented retinal cells. Light reflected from an object enters ommatidium and is converted to nerve impulses by receptor neurons in the retina. These travel through the optic nerve to be converted into a crude mosaic image by the brain.

C. CNIDARIAN MEDUSA (JELLYFISH), *Chironex fleckeri,* showing primitive ocelli. This is the exceedingly venomous Australian sea wasp. Two ocelli are visible between bases of tentacles along the edge of the bell. As the complexity of the phylum increases, primitive eye spots—simple accumulations of red, brown, or black pigment—evolve into more complex photoreceptors. The next evolutionary step is for the pigment spots to curve inward to make a cup. The most sophisticated cnidarian ocellus is a cup containing inverted retinal cells similar to the human eye.

D. HEAD OF CARIBBEAN PALOLO WORM, *Odontosyllis fulgurans,* showing its four elaborate ocelli. No image is formed and their only function is to discriminate between light and dark.

E. THE SCALLOP, *Aequipecten irradians,* has numerous ocelli peering out from undulations in the shell. Though sophisticated, they cannot form images. When the scallop senses danger (usually by smell), it swims away by clapping its valves (shells) together like castanets.

PLATE 31

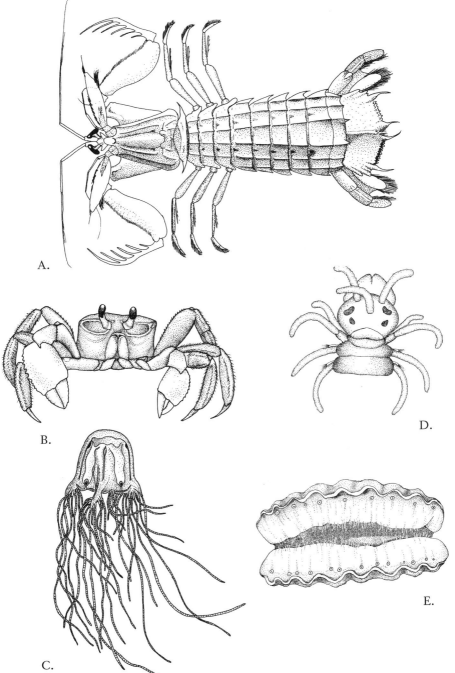

A.

B.

C.

D.

E.

Suddenly, my thumb feels like it has burst into flame. I drop rock and specimen. Blood flows copiously into the water. Groaning, I look at the bloody finger, realizing that I have just had an encounter with a "thumb buster," or mantis shrimp. These animals, looking like a cross between a praying mantis and a shrimp, are crustaceans of the order Stomatopodida.

Stomatopods are an unavoidable nuisance associated with collecting in shallow coral rubble. When undisturbed, they hide in a dark crevice in wait for an unwary minnow-sized fish or a true shrimp. When the animal is in range, the praying attitude changes, as if the prayer has come true. In a split second one of the supplicating front appendages reaches out in an uppercut, and a razor sharp "finger" slashes the prey in half. The stomatopod carefully moves out of its hole, its stalked eyes warily searching the environment for predators with the most sophisticated color vision in the animal kingdom.

The human eye has three kinds of cones—photoreceptors capable of picking up reds, yellows, and blues and melding them in the brain into a palette of thirty thousand hues. Mantis shrimps have as many as sixteen different color receptors, including three in the ultraviolet range. There are filters that polarize UV light and twelve truly color-sensitive kinds of cones. Instead of thirty thousand colors, the stomatopod eye can respond to millions of subtly different colors. Why? An answer proposed for the UV receptors is that many deep-sea stomatopods identify friend or foe by means of fluorescence (when one wavelength of light is absorbed and then emitted as another wavelength). In this case blue light changes into a glowing green UV-rich light. Potential mates are differentiated by their patterns of glowing spots, enemies by their patterns. This hypothesis deals with the UV sensitivity. But what accounts for the stomatopod's vastly extreme color perception? Nobody knows.

ᗰᗰᗰ Eyeball to Eyeball

As explained in chapter 2, the octopus is the "king of beasts." Its eyes are the culmination of invertebrate visual evolution. Its extraordinary ability to perceive colors is part of a unique nervous system. Images are processed almost instantaneously by the brain and, just as quickly, neurological messages are sent to tiny bags of red, yellow, and blue pigment in the skin called chromatophores. On command, they expand or contract to create subtle color combinations that meld into a replica of the octopus's surroundings. The change is as fast as the microsecond movements of the nerve impulses. The octopus perceives its environment and presto! It matches it. In all other animals, even chameleons, the changes occur hormonally and are slower by many orders of magnitude. When an octopus moves over the bottom, waves of color undulate across its body, matching its surroundings like colors flitting across a movie screen. It crawls over an algae bed—its skin is green. It enters the murky darkness of a cave—the green color darkens instantly. It becomes invisible.

When an octopus is crawling up the glass of an aquarium, it is possible to peer, eyeball to eyeball, into its eye. The pupil is a creepy, undulating black slot, the opening to its inscrutable personality. The octopus eye has an acuity rivaling ours. Examination of the inside of the eye reveals that although it creates refined images, it is as foreign to ours as is the compound eye. Its modus operandi has evolved along a different, molluscan, path. Our eye focuses by changing the shape of the lens. The eye of an octopus is cameralike. The lens moves back and forth to focus the image on the retina.

To stare into the eye of an octopus is to reveal its soul. It is the soul of a being so incomprehensible to us that it might as well be an alien from outer space.

The Facets of Knowledge

THE EARLIEST EVENTS described in this book (chapter 5) occurred at a marine laboratory off the coast of Washington state in July 1954. The most recent story (chapter 29) happened in a marine laboratory off the coast of Belize in July 2004, a period of fifty years. What have I learned about teaching in the half century between those events?

There is a conflict in the mind of every teacher as to how much "knowledge" must be sacrificed in order to make a course interesting. In the narrow sense, knowledge is how many facts can be crammed into the student's head. In the broad sense, knowledge is a composite of experiences: listening to lectures, touching objects, manipulating thoughts and things, and developing a gestalt of the subject that includes, in water-related biology, a feeling of muddiness, a briny sensation, the smell of fish and of rotten eggs (hydrogen sulfide released by a shovel from the anoxic layer of marsh mud). To provide the opportunity to face the reality of the sea and its habitats is a challenge that must be accepted. If the ocean is nearby, the students must literally be forced to change their lifestyles (spend a little less time in bars) to get wet in search of knowledge. Making them scour the beaches or stream banks on weekends to bring back specimens means that the instructor becomes the bad guy.

The mechanism I use to regain my reputation as a nice guy is to cancel one lab and convert it into the Invertebrate Zoology Fiesta, featuring paella composed of two phyla, four classes, and eight genera of invertebrates, in a sauce on Spanish rice (with saffron).

MENU
Three-Course Table d'Hôte Menu

Salad

Paella
Phyla Mollusca and Arthropoda

Mollusca: Classes Pelecypoda (clams, mussels, oysters), Cephalopoda (squids), Gastropoda (snails)
Arthropoda: Subphylum Crustacea, Class Malacostraca (shrimps, lobsters)

Quahog clams, *Mercenaria mercenaria*
Steamer clams, *Mya arenaria*
Mussels, *Mytilus edulis*
Squid, *Loligo pealii*
Shrimp, *Litopenaeus vannamei*
Oysters, *Crassostrea virginica*
Lobster, *Homarus americanus*
Scungilli (whelk snail), *Busycon caniculatum*

In appropriate sauce (some student in the class always
has a grandmother from Portugal or Spain).

Dessert
Choice of flan or homemade Toll House cookies

Note: we dig up many of the items on the menu.

All this is washed down with sangria made of a gallon of rotgut red wine diluted with quarts of orange juice so that I cannot be accused of getting my students drunk. All the while the halls of the biology building echo to the strains of flamenco music. We often have sixty or so students and faculty in attendance.

Thus, in addition to wallowing in mud collecting and identifying the requisite specimens, the student smells and tastes the invertebrates he or she is learning about. When the student cleans the food, he or she learns about the "beard" (byssal threads) of the mussel (*Mytilus*) and the outer plastic-like covering of all seashells (periostracum).

On all levels of science education in America, the student is expected to remain passive. Even the most well-intentioned teacher cannot help but concentrate on a litany of facts, circumscribing these facts with a veneer of methodology—at best telling the students how the scientist evolved his hypothesis and how he tested it. The student enters the university an intellectual blob, a tabula rasa upon which the instructor is expected to write "knowledge." The professor can recite stuff, write on the blackboard, use elaborate PowerPoint canned programs, present opportunities for virtual research in the form of videos of beating hearts (the beats of which vary with inputs of virtual substances).

But these lessons, at best, cannot simulate the gestalt of real sounds, smells, touches. It is the quest for the complete learning experience that led me to establish the Hofstra University Marine Laboratory in Jamaica. It was created to provide the opportunity (and inspiration) for professors to bring a class to the voluptuous world of the coral reef and tropical seashore. All the animals learned about in class are there, right in front of the students, staring back at them from behind a coral head, or on a rock, or among the roots of mangrove trees. There is nothing that makes me feel more like I am doing the right thing as when I say to the class, "Tonight we will wade out into the coral rubble to collect octopods."

But collecting mind-boggling animals is only part of the ultimate reality. Equally striking is the shock when a student sees a coral reef for the first time. It literally makes my spine tingle as this kid, who may have never even seen the ocean before, is transported by *me* to a tropical shore with talcum-white sand lined with dark green palm trees and edged by the turquoise sea containing incomprehensively vast biological riches. Providing the experience is exhausting for the teacher. But this reality far transcends the monotonous classroom lecture. The course I give in the Caribbean is extremely demanding on a scientific level—the course is only twelve days long—but at the end, virtually all of the students get good grades on both final exams, theory and reality (ecology and systematics), *because they are powerfully motivated by the experience.*

Thirty-five years ago I worked for the U.S. Office of Education and UNESCO on the design of an experiential science curriculum for the elementary school grades. The curriculum was infused by the teachings of the Swiss psychologist Jean Piaget. He postulated that each student reaches his or her own particular level of intellectual maturity at a different time, and the instructor must not assume that the class is a monolith of developmentally identical individuals. According to Montessori and others, contact with materials that engender the sought-after thoughts will stimulate lagging intellectual development. This elemental (and elementary) thinking infuses my teaching at the university level. If you want to evoke a thought process, allow the student to experience what it represents, and the thought will almost inevitably follow.

When reality is not available, striking—even shocking—examples of reality must serve in lieu. This book is a compendium of these examples. But every lurid lecture is boring when contrasted with walking the shores, smelling the smells, touching things, hearing the surf, or even better, donning snorkel and mask and descending under the sea to be one with the fishes, worms, and corals.

aboral — The side opposite the mouth of a radially symmetrical animal such as a cnidarian, ctenophore, or sea star.

adductor muscle — Large muscle that pulls together both shells of bivalve mollusks. Clams have two, and oysters and scallops one.

anti-BLB club — When a member of this club sees Bad Little Boys throwing rocks at a horseshoe crab, he or she is committed to throwing rocks at the BLBs.

anticoagulant — Substance that prevents clotting, as in saliva of a leech.

ambulacra — Five grooves in all echinoderms that contain tube feet.

amebocyte — An ameba-like cell found in the blood of most organisms. Some human white blood cells are amebocytes.

anoxic — Without oxygen. A zone of anoxic sediment lies beneath the muddy surface in marshes.

autotomy — When an animal deliberately breaks off a piece of its body. Sea stars and lizards commonly utilize this method of escaping from enemies.

axon — Long extension of a nerve cell (neuron) that carries the nerve impulse from the nerve cell body to the muscle or effector organ.

benthic — Bottom-dwelling.

biotic potential — Ability to produce numbers of offspring. Most schooling fishes have a high biotic potential, but top predators, like sharks, have a low biotic potential.

biological magnification — Prey organisms take in toxins. Predators eat them, further concentrating the toxin. By passing up the food chain, the toxin reaches a high level.

bloom — Sudden appearance of huge numbers of a species of an aquatic animal caused by the simultaneous maturing of the reproductive stage (not migration).

bommie — A huge isolated coral head usually encrusted with other organisms.

carnivore — Flesh-eater.

cephalization — Accumulation of sensory organs at the anterior of an organism, such as the human head.

cerata — Extensions of a nudibranch's upper surface containing branches of the gut. They can be ruffles, finger-like projections, or tufts.

chela — Claw of crabs, shrimps, lobsters, and other arthropods.

chelicerae — Mouth parts of spiders, ticks, and horseshoe crabs.

257

chitin, chitinous — The plastic-like surface of the shell of an arthropod. Composed of a fatty and a chalky layer, making it water repellent and hard.

chitinase — An enzyme that dissolves the chitin when arthropods molt.

chloroplasts — Green sac-like photosynthetic organelles inside plant cells and algae

chromatophores — Sacs of skin pigment used by animals to change their color. Found in squids, octopods, and other animals.

cilia — Microscopic motile hairs used for propulsion in ciliated protozoans and to remove particles from the human pharynx, coral polyps, and crinoids.

cirri — Anterior antenna-like projections from the heads of worms; also jointed appendages in commatulid crinoids that function as legs.

clasper — The reproductive structure of the male shark used to inseminate the female. It takes the place of a penis.

cloaca — Posterior chamber found in many animals, including frogs, sharks, and sea cucumbers, into which the reproductive, digestive, and urinary tubes empty.

commatulid crinoid — Modern motile crinoid; not stalked; a "feather star."

commensalism — Two species living together. One benefits, the other is not harmed.

compound eye — Arthropod eye made up of many photo-sensitive units (ommatidia).

copepod — A microcrustacean that often dominates the zooplankton.

cuticle — The plastic-like, impervious outer covering of many invertebrate phyla; in roundworms, for example.

cypris — Third developmental stage in some arthropod larvae. Used particularly in reference to barnacles.

cyst — Impervious, spherical chamber containing the larva in many phyla. Often microscopic.

detritus — Decomposing organic debris.

diapause — A period of minimal metabolic activity in arthropods and some mollusks. Similar to hibernation (winter) or aestivation (summer).

dinoflagellate — A protozoan with two flagella, one in a vertical girdle, one in a horizontal one. May contain chloroplasts and red eyespots. Sometimes toxic.

ecdysis — Molting, as in crustaceans, snakes.

ecdysone — Hormone facilitating molting in crustaceans.

endotoxin — Toxic substance produced by disintegration of harmful bacteria.

epitoky — The body of a worm develops a posterior modification for reproduction, the epitoke. It breaks off the parent worm and swims away, to burst and spill out its eggs or sperm. Found in the family Syllidae (palolo worms).

eyespot — Accumulation of photoreceptor cells. Used to differentiate light from dark in cnidarians, and other animals.

facultative mutualist — An animal in a mutualistic relationship that is not required for its survival. In the Caribbean, juvenile fishes often set up cleaning stations, but they eat other foods like zooplankton. As adults they are not cleaners.

flagellum — A whip-like structure used for movement, especially in flagellate protozoans and dinoflagellates.

gastropod — The largest class of mollusks. Examples are snails, nudibranchs, and sea slugs.

gonochoristic — Two separate sexes, male and female. (Previously called dioecious. This term is now relegated to plants.)

gonopore — Male or female reproductive opening. Used in this book to refer to the female opening.

great red-footed urchin — Large (to 12 inches), highly evolved subterranean sea urchin, *Plagiobrissus grandis*. Has distinctive red podia (possibly containing hemoglobin?).

hectocotylus — A structure on the third or fourth tentacle of a male squid or octopus used to inseminate the female with sperm packets. Breaks off in the mantle cavity of the female.

herbivore — Eats plants exclusively, like a cow.

hermaphrodite — Both sexes contained in same organism. Monoecious.

hormone — Minute amounts in blood affect the behavior of the whole organism or some of its parts. A means of internal communication. Produced by endocrine glands.

hypersaline — Unusually salty. Normal seawater is thirty-five parts per thousand of salt. Hypersaline water may be twice that. In the Dead Sea or the Great Salt Lake, or temporary ponds.

hypothesis — Based on previous observations, a scientist (or anyone) develops a possible explanation for a phenomenon and tests this explanation with an experiment. All experiments are tests of hypotheses. To examine something in a lab is not an experiment, even if the person is wearing a lab coat. Experiments require controls.

intertidal zone — The region between high and low tides; varies with the phase of the moon.

kentrogon — Post-larval stage of the parasitic barnacle *Sacculina*. Attacks crabs.

lateral line system — Fishes possess a linear sensory system along the flank capable of detecting vibrations in the water.

lunule — Oval holes in the tests (shells) of some sand dollar sea urchins, e.g., *Mellita*, that act as "spoilers," preventing waves from carrying them to shore.

mantle — A mantle wall is sometimes muscular (squid), sometimes diaphanous (clams). It encloses a chamber (mantle cavity) containing gills, gonad openings, and anus; found in all mollusks.

medusa — A jellyfish. The sexual stage in the life cycle of cnidarians.

microcrustaceans — Microscopic or near-microscopic crustaceans, e.g., *Daphnia*, copepods, and brine shrimps.

mutualism — A relationship between organisms where each benefits the other. A clownfish is protected by the nematocysts of its anemone host. It returns the favor by feeding the anemone pieces of animal matter too large for the clownfish to swallow.

nauplius — First larval stage of arthropods. It has one eye, swims with antennae, and does not feed.

nematocyst — A threadlike poisonous, sticky, or entrapping cell organelle used by cnidarians to trap food. "Poison dart." Produced by a cell called a cnidocyte.

nerve net — Primitive distribution network for nerve impulses in cnidarians and ctenophores.

neurotoxin — Toxin that kills by impairing the victim's nervous system. Death may come from paralysis of the diaphragm, causing suffocation.

niche, biological niche — The role of a species in its community, including the sum total of all biological factors affecting it. Includes climate, food, and predation.

nudibranch — "Naked gilled." A carnivorous, shell-less snail whose gills project from its surface. All other marine slug-like forms are herbivores.

obligate mutualist — An animal that must participate in a mutualistic relationship. It cannot survive without the relationship. In contrast to a facultative relationship, where the mutualist has other sources of sustenance and its partner can survive without its services.

ocellus — A primitive eye that consists of a group of photoreceptor cells, pigment cells, and, sometimes, a lens. Can be cup-shaped or spherical. Cannot form images.

ommatidia — The visual unit of the arthropod compound eye. Often hexagonal.

organelle — A structure inside a cell.

ossicles — Small bony structures such as those embedded in the skin of sea cucumbers. Bones in the mammalian middle ear.

oviducal gland — Produces nutritive secretion that surrounds the eggs in female squids and octopods.

parasitism — Relationship between two species where one is harmed and the other benefits. A mosquito is an ectoparasite that harms the host by feeding on its blood.

pheromone — A few molecules in air or water affect the behavior of another member of the species or, possibly, a closely related species. Often a sexual attractant.

photic ʒone — Depth to which light penetrates enough to make photosynthesis possible. Varies according to the transparency of the water. The more light-absorbing plankton in the water, the shallower the zone.

phototaxis — Inborn response to light exhibited by invertebrates and plants.

phytoplankton — The part of the plankton that is composed of chlorophyll-bearing, often-unicelled organisms such as diatoms. Smaller than herbivorous zooplankton that graze on it. Large floating seaweeds, like *Sargassum*, can also be considered phytoplankton.

plankton — "Wanderers." Any plant or animal incapable of swimming strongly enough to resist wind-driven currents. May be microscopic or large, such as jellyfish two feet in diameter. Composed of zooplankton and phytoplankton.

planula larva — "Hairy football." The microscopic ovoid, ciliated larva of cnidarians.

pleopod — "Swimmerets." Paddle-shaped appendages under the abdomen of shrimps, crayfish, and lobsters and their kin that flap back and forth. Eggs are attached to them to be oxygenated.

polyp — The sessile asexual stage of cnidarians. Alternates with medusae.

protandrous hermaphrodite — An organism containing both sexes sequentially, first male, then female. This requires morphological changes that take months or years.

protozoan — A one-celled organism that lacks chlorophyll. It can be motile (using cilia, flagella, or pseudopodia) or not.

protogynous hermaphrodite — First a female then a male. Not common, as it is advantageous for the small male to be first because sperm can more easily be stored than eggs.

radial symmetry — Body shaped like a pie. A line drawn through the center from any point on the circumference divides the organism into identical halves. No front or back. In animals, arms or tentacles can capture food from any direction.

radula — The file-like "tongue" of snails. It is actually a projection from the tongue-like odontophore.

red tide — Bloom of toxic dinoflagellates, *Karenia brevis* (= *Gymnodinium breve*). So abundant that their red eyespots tinge the water red.

regeneration — Replacement of a lost structure. Sea stars regenerate arms and sea cucumbers regenerate internal organs. Commonplace among echinoderms and lizards.

repugnatorial gland — Defense mechanism that makes some nudibranchs taste bad.

respiratory tree — Branched respiratory organ in the body cavity of a sea cucumber.

scutes — External bony plates protecting fishes, barnacles, and other animals.

sessile — Attached to substratum, unmoving. Examples are sponges and corals. May live on rocks or pilings. May move infrequently, as do sea anemones.

setae — Tiny hair-like spines on annelid worms and the hairs on the antennae of some crustaceans like brine shrimp.

sexual dimorphism — Having physical differences between the sexes. In fishes and invertebrates the females are often larger than the males and less brightly colored.

spermatophore — A sperm-containing capsule found in octopods, squids, and some crustaceans.

spiracle — Respiratory opening in nurse sharks, insects, and plant leaves.

stock enhancement — Increasing a population by adding aquacultured juveniles, like stocking a lake with trout.

stylet — Sharp, needle-like structure found in nudibranches, flatworms, and other animals.

subtidal zone — The region of the sea below the intertidal zone. Rarely exposed to the air. More uniform than the stressful intertidal zone.

swash zone — The part of the beach where a spent wave washes weakly upward, creating a turbulent region where frothy back-and-forth flow produces a stressful habitat.

symbiosis — Any of three trophic relationships between two species: parasitism, mutualism, or commensalism.

taxis — Inborn, automatic, unthinking behavior dictated by genes, as when a moth flies into a light (positive phototaxis)

temporary pond — A body of water that dries up for part of the year, or that may remain dry for many years and then reappear.

terrestrial niche — All the environmental factors affecting survival on land; see niche.

test — The "shell" of a sea urchin and other animals.

thigmotaxis — Inborn response to a surface. Tilapia juveniles press against the mother's flank and are protected (positive thigmotaxis).

trophic pyramid — A pyramid-shaped representation of food and energy relationships between levels of organisms, based mainly on their sources of nutrition. Shows that a large biomass of producers (plants) supports a smaller biomass of secondary and primary consumers (animals).

veliger — First larval stage of mollusks; has a wing-like, ciliated velum; swims. Settles on the bottom after this motile stage.

X gland — In eyestalk of crustaceans. Produces a hormone that inhibits molting.

Y gland — Not in eyestalk; promotes molting.

zoea — Second larval stage of crustaceans (after nauplius). Has two eyes and is predaceous.

zooplankton — Predaceous animals that feed on phytoplankton. May eat herbivorous zooplankton. Range in size from jellyfish to zoea larvae, protozoa, larval fishes, and worms.

zooxanthellae — Chlorophyll-containing mutualistic dinoflagellates living in tissues of corals, jellyfishes, giant clams (*Tridacna*), and other animals. Contribute food to the host.

THE DRAWINGS in this book are a collaborative effort by Susan Kaplan and Sandy Chichester Rivkin. Theirs was "a collaboration made in heaven" reflecting the merging of their extraordinary talents into a seamless (to my eyes) creation of thirty-one esthetically and scientifically valid plates. I cannot express my gratitude enough in this acknowledgment of their efforts.

Most of the drawings are composites, taken from books, the internet, and in a few instances, the animal itself. There are a few "classical" pictures that have been redrawn to be more or less identical to their sources. These will be listed below.

Houghton Mifflin has kindly consented to allow us to use drawings made by Susan Kaplan that were taken directly from the plates in each of my field guides. These are:

frontispiece; Pl. 11d, e; 24d, e: Eugene H. Kaplan, 1982, *A Field Guide to Coral Reefs of the Caribbean and Florida (Peterson)*, Houghton Mifflin Co., Boston

frontispiece; Pl. 7c; 15e; 17b; 23d, e; 24b, e; 27c, f; 31a: Eugene H. Kaplan, 1988, *A Field Guide to Southeastern and Caribbean Seashores (Peterson)*, Houghton Mifflin Co., Boston

Redrawn from the following books:

Pl. 1a; 20e, f: Vicki Pearce, J. Pierce, M. Buchsbaum, R. Buchsbaum, 1941, 1987, *Living Invertebrates*, Blackwell Scientific Publications, Palo Alto, CA

Pl. 30b, d: Mark Cowardine, K. Watterson, 2002, *The Shark Watcher's Handbook*, Princeton University Press, Princeton, NJ

Pl. 12e: Tijs Goldschmidt, *Darwin's Dreampond*, 1997, The MIT Press, Cambridge, MA

Pl. 3d: Stephen Spotte, 2002, *Candiru: Life and Legend of the Bloodsucking Catfishes*, Creative Arts Book Company, Berkeley, CA

Pl. 8d: Ralph Buchsbaum, M. Buchsbaum, J. Pearse, V. Pearse, 1987, *Animals Without Backbones*, University of Chicago Press, Chicago, IL

Pl. 17a: Jocelyn Crane, 1975, *Fiddler Crabs of the World*, Princeton University Press, Princeton, NJ

Pl. 1e: W. D. Russell-Hunter, 1970, *A Life of Invertebrates*, Macmillan Publishing Co, New York, NY

Pl. 8a: Jan Pechenik, 2000, *Biology of the Invertebrates*, McGraw Hill, New York, NY

Page numbers in **boldface** refer to plates.